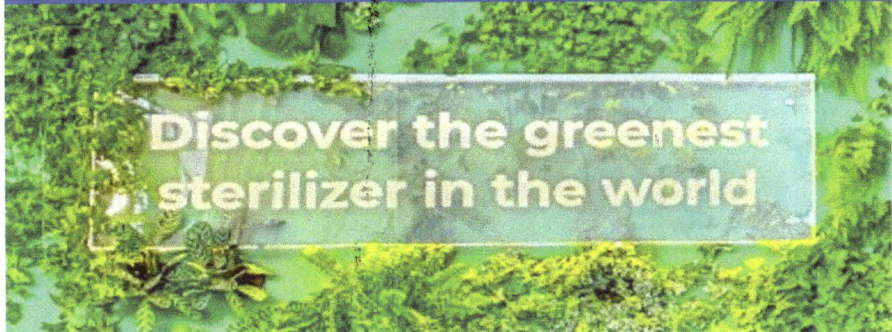

LEADING BEYOND CHEMISTRY

TO IMPROVE LIFE, TODAY AND TOMORROW

Innovation beats tradition

Innovation statt Tradition

PROCESS. HEAT. TECHNOLOGY

FOR EFFICIENT AND SUSTAINABLE HEAT AND POWER APPLICATIONS

Henk Akse

Sustainable Manufacturing Processes

Also of Interest

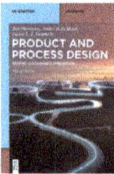

Product and Process Design.
Driving Sustainable Innovation
Harmsen, de Haan, Swinkels, 2024
ISBN 978-3-11-078206-6, e-ISBN 978-3-11-078212-7

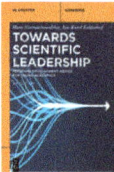

Towards Scientific Leadership.
Personal Development Advice for Young Academics
Niemantsverdriet, Felderhof, 2024
ISBN 978-3-11-132531-6, e-ISBN 978-3-11-132564-4

Sustainable Process Engineering
Szekely, 2024
ISBN 978-3-11-102815-6, e-ISBN 978-3-11-102816-3

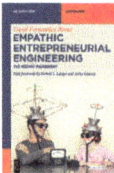

Empathic Entrepreneurial Engineering.
The Missing Ingredient
Fernandez Rivas, 2022
ISBN 978-3-11-074662-4, e-ISBN 978-3-11-074682-2

Process Intensification.
Breakthrough in Design, Industrial Innovation Practices, and Education
Harmsen, Verkerk, 2020
ISBN 978-3-11-065734-0, e-ISBN 978-3-11-065735-7

Henk Akse

Sustainable Manufacturing Processes

———

The Coaching Method Enabling Companies to Innovate

DE GRUYTER

Author

Ir. Henk Akse
Traxxys Innovation & Sustainability
Adenauerlaan 17
3446 HS Woerden
The Netherlands

ISBN 978-3-11-138342-2
e-ISBN (PDF) 978-3-11-138366-8
e-ISBN (EPUB) 978-3-11-138380-4

Library of Congress Control Number: 2024945035

Bibliographic information published by the Deutsche Nationalbibliothek
The Deutsche Nationalbibliothek lists this publication in the Deutsche Nationalbibliografie;
detailed bibliographic data are available on the Internet at http://dnb.dnb.de.

© 2025 Walter de Gruyter GmbH, Berlin/Boston
Cover image: peterschreiber.media/iStock/Getty Images Plus
Typesetting: Integra Software Services Pvt. Ltd.

www.degruyter.com
Questions about General Product Safety Regulation:
productsafety@degruyterbrill.com

Preface

"How can it be, so many sustainable technological solutions already exist and so little companies in the process industries actually deploy them?"

While advising companies in the chemical sector since 2004 as an independent consultant on innovation and sustainability, people have asked me this question over and over again.

This empirical guide answers the question. It assumes that you have a leading role inside your company and you have the ambition to identify and deploy an innovation in your existing chemical manufacturing process to improve process sustainability.

You will notice that the guide has a four-stage approach. First, it assists you in a situational analysis of your company. Depending on the outcome, it advises on a course of action with added value to the company as the prime criterion. Second, it gives you practical definitions of innovation and sustainability you can work with. Third, it offers an empirical method that enables you to innovate the process yourself toward a higher level of sustainability. Fourth, it gives you a set of Sustainable Design Guidelines to make sure you integrate your innovation in a sustainable way.

Hoping and trusting this guide is a useful and inspiring read!

January 2025
Henk Akse
Traxxys Innovation & Sustainability
Woerden
The Netherlands

https://doi.org/10.1515/9783111383668-202

Foreword

The *importance of sustainable manufacturing processes*, required to produce the global needs and within the limitations provided by the planetary boundaries, cannot be overstated. The relevance of a book on *a coaching method enabling companies to innovate* and to adopt these sustainable manufacturing processes is therefore very timely, and this is precisely what the aim of this book. But this book is more than that. It is a practical guide based on the vast consulting experience of the author Henk Akse with manufacturing companies in their struggle to innovate and introduce sustainable intensified production processes. It elaborates on a novel approach – the Self-Innovation-Method, 6-steps SIM – in which essential ingredients to be a successful innovative company are identified. This ranges from operational experience and imaging skill to a working knowledge of physics and chemistry and a broad technology overview. In particular, it is the workforce in a chemical plant – the people from work floor up to management – that determines the success of an innovative transition within a company. Not so much the innovation method is key. Key is the team of skilled people having the attitude to adopt change to achieve their sustainable goals.

But the book is more than a primer on this novel approach of a coaching method enabling companies to innovate. It is a book which is based on several lifetime experiences. Anecdotes are introduced which illustrate a particular strategy to disruptively get the companies rethink what their original aim is and why they ask for consultation in the first place. References to scientific articles explaining why 6-steps SIM works are presented in distinct blue boxes. All in all the book reads as a novel in which the author introduces you to his consulting practice. He does not hesitate to share his failures as well as his successes. It becomes clear beyond doubt that his lessons learned from this lifelong journey have resulted in a successful innovative method to introduce and adopt sustainable manufacturing processes in industry.

I was particularly struck by the observation of the fact that many larger, well-established companies are less willing, or able, to change. It seems that innovation and change must come from newcomers. This is an important observation, especially when we look at the challenges that the energy-intensive (chemical) industries are facing, such as, e.g., defossilization. Closely related to this is the introduction of the Circular Innovation cycle. This cycle distinguishes itself from the usual Linear Innovation stage gate approach. From my own experience, the Circular Innovation cycle is essential. This is especially true if we look at the challenges of the energy and climate transitions. These transitions require innovations in almost every aspect of the production chain. Only the system integration looking on the whole chain of interdependent steps can lead to real innovative and sustainable manufacturing process. In other words, the needs and limitations need to be taken into account while innovating.

https://doi.org/10.1515/9783111383668-203

For the above-mentioned reasons, I think the book has some important novel and innovative contributions and will inspire you to think innovation on a deeper level, and how the various actors can act together to achieve a successful introduction of truly innovative and sustainable manufacturing processes.

Prof.dr.ir. M. C. M. (Richard) van de Sanden
Scientific Director
Eindhoven Institute for Renewable Energy Systems (EIRES)
The Netherlands
Professor
Department of Applied Physics
Eindhoven University of Technology
The Netherlands

Tilburg, January 2025

Acknowledgments

Reading a book is a trivial thing. Writing one is exactly the opposite. Message and content do not drop from the sky on a sunny afternoon. For me, preparing this book appears to have been a process that has continued for many years without me knowing what was going on in my subconscious.

What I do know is the fact that there have been people, companies and institutes in the past two decades that have contributed to the "Werdegang" of this guide. Here I like to thank them.

First of all, I thank Jacqueline Cramer – former Dutch Minister of Housing, Spatial Planning and Environment, emeritus professor in sustainable innovation and currently transition broker – for convincing me to start my consultancy company. She did this in the most charming, challenging and inescapable way.

Next and equally important, I thank each of my clients for trusting me in a joint effort to improve their process sustainability. They have given me space and time to experiment, learn and improve.

I thank Bart Drinkenburg – former Director of Corporate Technology at DSM and late professor in chemical engineering – who showed me how to have the courage to challenge established designs and to introduce creativity in manufacturing environments where strict regulations rule the waves. I thank Bart very much for leading by example: first as teacher and later as colleague!

I thank Henk van den Berg – formerly employed at DOW Chemical in process design and currently professor in process plant design – for keeping me on the methodical track and for many years of fruitful cooperation. I thank Tony Kiss – former employee at AkzoNobel's process design and currently professor in process systems engineering – for publishing an increasing series of creative sustainable process designs and for sharing many of these with me. They are outstanding, inspiring and leading examples of sustainable manufacturing processes.

Richard van de Sanden – former director of the Dutch Institute for Fundamental Energy Research and currently scientific director of the Eindhoven Institute for Renewable Energy Systems – deserves a special word of thanks: First, because he trusted me in introducing chemical reactor engineering principles into the universe of plasma physics. Second, because he and his coworkers have made it unintentionally clear – much to my dismay – that the real process innovators of this world are physicists and not chemical engineers. We know our place now.

Kees Vendrik was former Member of the Dutch House of Representatives and is currently chairman of the National Climate Platform. I thank him for making me aware of two things in the brief discussions we had. He stresses the urgency of the energy transition and shows the importance of organizing the transition from bottom to top. This bottom-up concept has made me writing the guide for people in the work-

https://doi.org/10.1515/9783111383668-204

place of companies. Finally, I thank Karin Sora – vice president at the Gruyter Brill Publishers – for swiftly recognizing the fact that all ingredients for a book were ready in my mind and for asking me to write it. I thank content editor, Ria Sengbusch, for the useful and enjoyable sparring sessions we had!

January 2025
Henk Akse

Author's biography

Henk Akse worked half of his career for different companies, which all focused on energy and environment (Shell, Paques, Vattenfall and Arcadis) before founding his own firm Traxxys in the fall of 2004. Traxxys is an independent consultancy company focusing on innovation and sustainability.

At Shell, Henk was responsible for innovating existing chemical manufacturing plants (Shell International Chemical Company) and for developing novel oil additive processes (Shell Research ETCA), thus delivering input for the intended construction of two oil additive plants in Berre (France) and Stanlow (UK). He managed international teams focusing on the design of next-generation zero-waste chemical plants for ethylene, propylene and butylene (Shell, The Hague, SICM) by supervising engineering projects in cooperation with companies like Stone & Webster (Houston), Snamprogetti (New York), Honeywell UOP (Des Plaines Illinois) and Lummus ABB (The Hague).

In 1998, Henk founded the Dutch Knowledge Network about process intensification (PI). PI holds the promise of bringing down capex and enhancing process circularity while keeping plant capacity at level. Today, SPIN-NL (Sustainable Process Intensification Network NL) is a thriving and inspiring network attracting young chemical engineering professionals and encouraging them to integrate sustainability in their daily work.

Between 2004 and now, Traxxys has advised over 65 different chemical companies, R&D institutes and governmental organizations in the Netherlands, Singapore, the United Kingdom, Sweden, Belgium and Ohio (USA). Traxxys has developed an empirical innovation method called 6-steps SIM in cooperation with clients. The essence of the method is that it is the company itself that innovates. Its first deployment was in 2014.

Henk has a partner, two daughters and a grandson. He participates in Coastal Rowing Competitions (Royal Netherlands Yacht Club) and in cycling tours (Elfstedentocht Friesland). His main other areas of interest comprise photography and visual arts. He is an enthusiastic visitor of the annual Bach Academy in Bruges (Belgium) and of "Concerts en nos Villages" in the surroundings of Poitiers (France).

https://doi.org/10.1515/9783111383668-205

Contents

How to position this book

i

I know I can do it, but I don't know how I do it

Keith Richards, lead guitarist of The Rolling Stones, in an interview with Jimmy Fallon, New York, November 2023

This book claims to be a practical guide with the purpose to assist you in innovating existing and novel manufacturing processes involving chemical conversions toward a higher level of sustainability. It assumes you have a leading role in your organization. The Self-Innovation Method (6-steps SIM) presented in this book is strictly empirical. It enables you to innovate your manufacturing process toward a higher level of sustainability yourself, if desired, coached along the way.

As such, the guide occupies a modest yet unique position in the wide array of books on innovation emerging from corporate multinationals to academia, most of which are founded on scientifically validated bases.

If deployment of innovation is the target, it is clear beyond doubt that human behavior, human interaction and value and mindsets of people inside your company are equally important as the availability of a method.

This is why the scope includes chapters that help you to be effective in initiating innovation by understanding your company's Narrative comprising vision, mission and identity, your colleagues' behavior, internal and external company-affecting factors and by understanding what your company really is about.

The book claims to enable the achievement of two conditions that have to be met before you can embark on the 6-steps SIM. These are a Company Support Base for your proposals and a Company Innovation Team that has the mindset and the enthusiasm to make it happen.

Finally, although this guide is purely empirical, it contains concise trips to science to explain why things work. You will find these explainers in blue boxes throughout the book. Because, rephrasing Mr. Richards of the Greatest Rock and Roll Band in the World: "I know it works, but I don't know how it works."

Henk Akse

Disclaimer: Empirics, findings and advice in this book stem from the consulting practice of Traxxys during November 2004 until the present day. Process details of client companies are not given in respect of prevailing Non-Disclosure Agreements. Examples from consulting practice are given in such a way that backtracking to individual companies is avoided.

https://doi.org/10.1515/9783111383668-207

1 You and the company

Abstract: The chapter explains for whom this book is written. It analyzes and describes the context in which your company is operating. Starting with analyzing the Narrative of the company (vision, mission and identity), the chapter enlightens the purpose and meaning of the company structure, internal and external factors affecting the course of the company. Subsequently, sustainability aspects of the Narrative and the existence of deployed sustainable projects are analyzed. Four different starting situations are identified. Four strategies are unfolded for developing a sustainability project that will add value to the company depending on the situation you find yourself in.

1.1 Getting acquainted/for whom is this guide?

Before diving into the topic of improving the sustainability of manufacturing processes, I like to get to know you a bit better. Who are you as a person? What is your profession? What is your drive? Do you have a working knowledge relevant for your job? What is your skill set? Do you have a network in your organization? Do you have a sound understanding of what your company is really about? Who are the actual decision-makers in your company? What is their drive? And – most importantly – do you know what you want to achieve in your job?

A brief introduction of myself: Being a chemical engineer by training, I kicked off in 1984 with a strong drive to improve environmental, energy and resource aspects of the chemical process industry at large. Now that is a broad target. Soon I discovered that drive helps, but it is not the jack-of-all-trades that implements the lasting change. In those early years of my career, I was looking for answers to the same questions I am asking you now.

I have written this guide assuming you have a leading role in your company with the ambition to improve the sustainability of your company's manufacturing process. That being said, the guide's scope in terms of company type is broad. Your company may be a startup with one or two colleagues on the payroll. It may be a scale-up with its first successful sales in the market. It may be a small to medium enterprise with an established market position. It may be a family-owned private company with well over 100 years of existence. It may be a multinational with worldwide presence. The guide's scope in terms of process type is limited to manufacturing routes comprising all presently known commercial chemical and physical unit operations. The guide assumes you have access to technologically trained academic and operational people with experience in and knowledge of plant operation, chemistry, chemical engineering, physics, catalysis, reactor engineering, separation technologies, process design, upscaling, cost estimating, process economics, process improvement and troubleshooting of chemical processes.

https://doi.org/10.1515/9783111383668-001

This guide will help you in asking meaningful questions at any stage of the innovation process and in finding useful and effective answers to those questions. You will become successful in innovating your process toward a higher degree of sustainability. As a matter of fact, the dialogue I am having with you throughout this book is similar to the way I have been consulting companies since 2004. The right question yields the right answer. The better you become in identifying the right questions and in daring to ask them, the more successful you will be.

1.2 Context

Is it beating around the bush – starting this guide writing about context instead of giving you a proven set of rules that help you in progressing toward a more sustainable manufacturing process? Certainly not! Let me explain.

First of all, what is context? In this guide, I consider the context to be twofold. The first layer of context is everything that happens inside your company: Interaction with colleagues, managers, team leads, departments, human resources, the manufacturing plant, internal courses and safety training; contributing to strategy sessions with the CEO, including presentations about vision and company's Narrative, quarterly profit and loss meetings, daily talks with plant staff; annual performance sessions with your boss, interesting talks with plant operators, weekly internal seminars for all departments, unexpected plant shutdowns; and many others. Need I give you more? It feels like an arbitrary mix of relevant and irrelevant, of urgent and nonurgent matters. Actually, it is. A working environment is not a clear-cut straight line from where you are to your final project target. At best, it is a flow that constantly changes the speed and direction and you better learn to surf it!

The second layer of context is everything outside the fence that affects your company's business. Impatient customers in need of faster delivery, roadblocks by demonstrators, Health, Safety and Environment (HSE) directives from local governments, caps on CO_2 emissions, prices of energy and feedstocks, negotiations with labor unions to boost wages and improve working conditions, mergers and acquisitions of other companies including due diligence exercises at the remote site, informative sessions with neighbors living in the immediate surroundings of your plant, markets requiring new products or more stringent specs on existing ones.

My purpose is to make you aware of a few factors that are generally valid for any company. These factors you need to understand and to take into account when you set off with your ambition to pursue sustainability in your company. Of course, the complexity of a multinational differs from a three-person startup. Let me mention what I consider to be important factors.

1.3 Company's Narrative

Let's choose your company's Narrative as a starting point to help you understand the situation with respect to sustainability in your company. A Narrative in my view is an inspiring piece of text describing vision, mission and identity of the company. The vision is about what the company wants to be in 10–20 years from now. The mission describes what it wants to achieve now. The identity describes the culture inside the company: how people work together, how they achieve results and what are the do's and don'ts. Most of company's Narratives I have come across since 1984 – stories told to employees and to the outside world explaining what the company is about – are solid pieces of work, logically structured, with transparent content, clear vision, mission and identity, compliant with state-of-the-art practices of management, marketing and sales, QHSE, logistics, etc.

Narratives often intend to convey a value set to the world outside as well. It may look like this. Our company adds value to its customers by manufacturing chemicals following the highest standards of safety, quality and cost. Our colleagues operate in an open atmosphere. We embrace the behavior showing loyalty, trust, honesty and a critical yet positive attitude toward each other and toward results. We are supportive toward each other and we conduct our jobs in a professional and ethical manner.[1]

However, walking through the corridors of your office, talking to colleagues, listening to what they do often will make you aware of the fact, what is actually happening is not fully matching the Narrative. Everywhere you can find new initiatives, projects and practices not yet covered by the Narrative. Reversely, you may also be looking for parts of the Narrative that do not seem to have a tangible counterpart inside the fence.

My message is, organizations are alive just like the people who work in them. Vision and mission may look to be cast in concrete, but in fact they are in constant flux. Sometimes the time constants of these changes are very large, which may leave you with the impression that nothing changes. The relevance of this for you is that there may be an opportunity here. You may change a Narrative not yet covering sustainability in one that is. Alternatively, you may be able to strengthen a Narrative supporting sustainability in text only by developing and executing tangible circular projects.

---

1 Scientific evidence confirms the existence a Narrative is of great importance, especially when an organization is in transition. A Narrative not only connecting the old situation to the new desired situation but also presenting the underlying value set is most effective [1].

1.4 Purpose and implications of structure

In addition to Narrative, vision and mission, mid- and large-sized companies often show well-established organizational structures. Structure – often visualized in Org-Charts – has purpose and meaning.

Its purpose is to realize the company's objectives – e.g., manufacturing and selling chemicals – in order to maximize the effectiveness of the workforce toward this objective. Mintzberg's classic [2] explaining why organizations need a certain structure illustrates this crisp and clear. Structure may vary significantly from one company to the next: from completely flat – a type favored by small start/scale-ups – to multilayered hierarchical, as often seen in large corporates. Regardless of size, a good match between goals and structure results in a well-managed cost-effective organization. Even then, there will be roadblocks, which we will discuss in Section 2.4.

On a concrete level, the organizational structure supports a set of business functions and links them in a logical way: Operations, Sales, Logistics, Technology, Utilities, Staff, R&D, Legal, Human Resources and Marketing. Each of these functions reports to general management.

This yields two important implications. First, it is the combination of functions that is essential in running the company to add value in the market place. As much as you would like to believe otherwise, the technology department is but one factor in the total game. A proposal can be brilliant from a technological point of view. If it makes no sense businesswise, it is not going to fly.

The second implication is that every sustainability initiative you propose will have to be endorsed by a limited group of three to ten people heading these functions – or departments, depending on the size of your company.

Finally, the existing structure of your company may leave you with the impression that everything is fixed. Please keep in mind that structure follows function. If you can convince your management that your sustainability initiative adds the lasting value to the company, this will eventually be translated into additional functionality of the organization. The added functionality will subsequently be supported by additional company structure.

1.5 The world outside

Considering the title of this guide perhaps the most topical thing that springs to mind when contemplating how the outside world affects your company might be regulations concerning CO_2 and NO_x emissions of your manufacturing process. Emission targets differ from country to country. In general, the trend is downward. If your company emits these gases, targets will impact your company's business. Your projects may anticipate these targets enabling the company to deal with them timely and effectively.

In the past, there have been numerous other external factors significantly affecting companies. I would like to illustrate this with a few out of many examples.

Total Quality Management (TQM) already kicked off in the 1920s. In the decades that followed, people like Shewhart, Deming, Juran, Feigenbaum, Crosby and Ishikawa made large contributions to the discipline [3]. It made its entry into the process industries in the 1980s and 1990s. TQM has had a profound impact on the way manufacturing plants are operated throughout the world. Your project proposals will need to meet quality criteria.

HSE has moved up on the agenda of companies in the process industries since the 1970s and will continue to be high-ranking meeting items. Health of employees has to be guaranteed. Plants have to be safe. Environmental emissions, toxic or otherwise, should be absent. HSE has become the foundation of any company's License to Operate. Each of these factors affects every initiative you come up with. Sustainability projects need to meet criteria emerging from HSE. Nowadays, Quality and HSE have merged into QHSE.

Since over a decade, compliance has entered the business arena. Triggered by huge scandals in the seemingly untouchable world of "Haute Finance" of Wall Street in 2008 followed by the collapse of the US financial system as it used to be, compliance has made it way to other business segments as well. Today it is a steady and recurring factor in business life, not only in banks and insurance companies, where hundreds of people exclusively deal with the question of whether a financial transaction complies with the rules set by the National Bank and the Financial Markets Authority but also in industries like the one you and I are discussing here.

Another trend that in all likelihood will have reached your company is benchmarking [4, 5]. An example of a well-known benchmarking company that kicked off consulting the fossil-based industry worldwide is Solomon. Since 1980, their methodology has enabled clients to understand how their operations compare against peers in key operational areas. Solomon now also produces sustainability benchmarks. The upshot is that your company is informed about its sustainability performance (expressed in suitable Key Performance Indicators) compared to your peer companies. We will get back to this in Section 2.7.

A wider perspective on how top-tier consultancy companies deal with sustainability can be found in Explainer 1.

Explainer 1: Top-tier consultancy companies and sustainability.

No.	Company	Focus areas	Remarkable	References
1	Solomon	Benchmarking Sustainability	It is a sustainability service delivered by a company that traditionally serves fossil-based industries	[4–7]
2	McKinsey	Hyperscaling green business: $275 trillion investment is needed between now and 2054 to become net zero in 2050 Creating value with decarbonization Amplifying climate investment Accelerating deployment of net-zero technologies Scale nature-positive business Collaborate for global impact	It is about an integral approach – *"never just tech"*	[8]
3	IBM	Use IT as a key driver to reach sustainability results Use data as a catalyst for transformation Prioritize energy and emissions management Align business objectives with improved environmental outcomes Use AI to speed things up Ensure broad "C-level involvement": this means the buy-in of a widening group of decision-makers is required Activate sustainability across the organization	It is about the CFO who plays a key role in developing a sustainability strategy	[9]
4	Deloitte	Accelerating energy transition from investment to implementing innovation Developing sustainable food systems Building a circular world	It is about going from a linear to a circular economy	[10]

5	Ernst and Young	Deploy "value-led" sustainability: business creates value for sustainability and sustainability creates value for business Use AI Understand the climate risk to move you from ambition to action Reframe the strategy to enable sustainability to deliver long-term value Accelerate transition Adapt governance so that value sustains over the long term Build trust with stakeholders, regulators and investors	It is about resilience and sustainability	[11]
6	KPMG	Achieve net-zero carbon emissions by 2030 Give financial markets, clients and our leaders clear, comprehensive, high-quality information on the impacts of climate change Understanding and improving our impact on nature and biodiversity	It is about a more sustainable and resilient future	[12]
7	PwC	Climate risk, resilience and adaptation Impact management for sustainable business strategy Net-zero transformation Sustainable capital Sustainability reporting	It is about resilience PwC explicitly builds business cases to engage with UN Sustainability Development Goals	[13]

Let us look at a few observations following from Explainer 1. Of these top seven consultancies, Solomon, McKinsey and IBM each take a unique view on sustainability, whereas Deloitte, Ernst and Young, KPMG and PwC more or less share a similar view. Solomon treats sustainability as a company reality on par with traditional items like operational cost and energy use. McKinsey stresses an integral business approach in which sustainability is but one element. IBM gives the CFO a central role in development of a sustainability strategy. The other consultancies focus on enhancing circularity and resilience of manufacturing processes and businesses.

The most recent external factor affecting your company may well be the Corporate Sustainability Reporting Directive (CSRD) issued by the European Union. It has come into effect in 2024 [14, 62]. This law requires European companies to issue an annual report about the impact of their activities on man and environment. I suspect this will increase pressure on companies to ensure all business and manufacturing procedures related to sustainability are compliant with specific standards and laws issued by the EU.

In hindsight, some of the mentioned external factors that came into fashion have proven to be of lasting benefit to companies. Others have appeared to be hampering entrepreneurship, often leading to large amounts of admin. Perhaps it is a consolation that not all external trends imposed upon companies have a lasting impact. Table 1 shows 100 management fads and fashions since the Second World War up to the year 2000 [15].

Table 1: One hundred management fads and fashions since the Second World War (1945–2000).

Acceptable Risk	Distributed Intelligence	Knowledge Management	Servant Leadership
Assessment Centers	Downsizing or Rightsizing	Learn Manufacturing	Social Responsibility
Automatic Factories	Diversification	Learning Organizations	Spin-Offs (Divestiture)
Baldridge Award	Diversity Training	One-minute Management	Stewardship
Balanced Scoreboard	Dress-Down (Casual) Friday	Organization Development	Strategic Planning Units
Benchmarking	Education Initiatives	Out-of-Box Thinking	Subcontracting
Broad-Banding	Electronic Data Processing	Outsourcing	Supply Chain Management
Business Ethics	Emotional Intelligence	Managed Health Care	Takeovers
Business School Offerings	Empowerment	Management by Objectives	Team Building
Cafeteria Programs	Ethical Leadership	Mgt. by Walking Around	T-groups
Centralization	Excellence	Matrix Management	Theory Z (and Theory X & Y)*
Change/Creative Destruction	Experience Curve	Myers-Briggs Type Indicator	Time-Based Competition
Chaordic Organizations	Flat Organizations	Pay for Performance	Time-Motion Studies
Computerization	Flex Time	Portfolio Management	Time Sharing
Complexity	Free Information Exchange	Post-Capitalism/Co-opetition	Total Quality Management
Computer-Integrated Manufacturing	Functional Teams	PERT (Prog. Eval. and Rev. Tech.)	Training
Conglomeration	Internet	Project Management	Transactional Analysis
Convergence	Intrapreneuring	Privatization	Transformational Leadership
Core Competencies	ISO-9000	Quality Circles	Value-Based Management
Corporate Culture	Issues Management	Reengineering	Value Chain Analysis
Critical Path Analysis	Japanese Management	Restructuring	Virtualization
Customer Driven	Job Enrichment	Sales Force Automation	Zero-based Budgeting
Data Warehousing	Job Sharing	Scientific Management	Zero Defects
Decentralization	Joint Ventures	Self-Managed Teams	Zero-Latency Enterprises
Demassing	Just-In-Time	Sensitivity Training	et cetera*

*Note: et cetera occupies position 100. The table's title can be made honest by listing Theory Z and Theory X and Y separately.

How are these external factors relevant to you? Well, for one, those trends that survive the "Fads and Fashions" phase are the ones that stay around. You will have to deal with them, since they will affect your proposals.

Second, the surviving ones could hamper your innovation initiative. Take for example "Compliance." It is not in Table 1 since it popped up after 2000. It means: "Abiding by a set of rules. For your business to function legally, it needs to comply with specific industry standards, laws, regulations, and ethical conduct standards that apply to your business." This does not immediately breathe the thrilling atmosphere of innovation. For a start, a good working relationship with your Compliance Officer may work wonders. But when compliance rules the waves, your real challenge is to create an innovative atmosphere in the Team you are going to work with. Section 2.5 gives you a set of tools you can apply to make this happen. Guaranteed.

1.6 The discrepancy factor

This factor may well be the least discussed and the most determinative one to take into consideration when you want to take on a sustainability project. It seems trivial, but it is not when you look at it more closely. Often, this factor is one of the root causes of continuous confusion about the direction of the organization. It concerns the gap that may exist between the company's Narrative and everyday business life. Figure 1 summarizes four dimensions of this factor limited to the subject of sustainability.

Like any decent consultant I am here to translate your complex world into four simple situations! If you find yourself in Situation 1, your company's Narrative says sustainability is at the core of its "raison d'être." Lo and behold – looking around, you discover deployed sustainability projects! You are blessed. The Narrative and the actual situation in the company are consistent. Although there is no discrepancy here, this does not mean you can sit back and relax. Projects do not initiate themselves. When you want to develop a sustainable project yourself, you still need to effectively manage all of the factors we are discussing in this chapter before you can take a deep dive into its interesting and challenging technological aspects.

Situation 2 is of a totally different nature. Your company's Narrative tells you, your colleagues and the outside world that sustainability is part of its value set. Yet, despite thorough scrutiny, you cannot find anything that even remotely resembles deployed circularity or sustainability projects. A simple count reveals that about 80% of the companies I had the honor of consulting over the past two decades are in this phase. Here we do have a discrepancy between text and action. These companies talk the talk. They seriously want to achieve something in this area. The availability of budget proves this. Management puts people and money in task forces that screen the existing manufacturing routes for opportunities to enhance process sustainability. Management hires outside experts to coach inside Teams. From my consultancy practice, one company comes to mind that excels in situation 2. Employees at different levels in this company broadcast

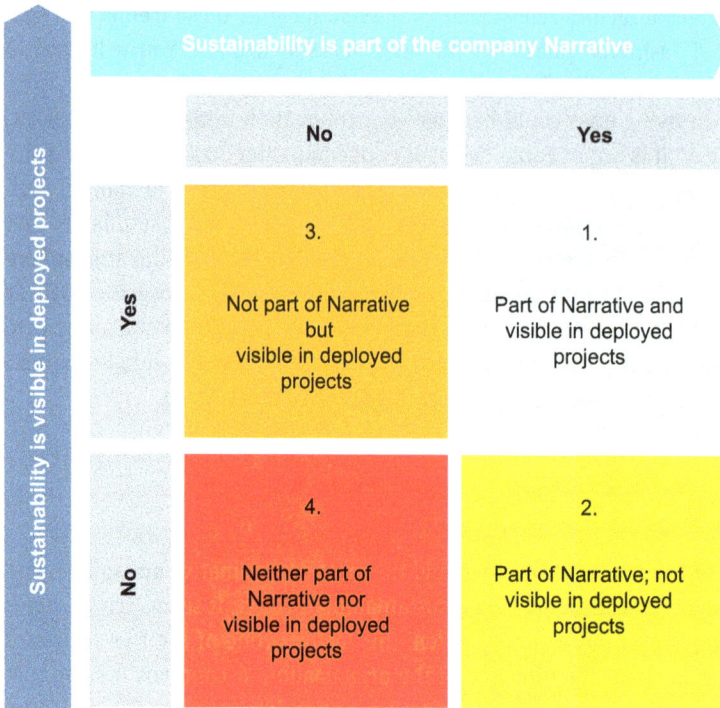

Figure 1: Company's Narrative and deployed sustainability projects.
©H.N. Akse 2025

completely different signals to the outside world. On board level one of the members is CSO (Corporate Sustainability Officer). She is exclusively busy with all sorts of programs and techniques that measure company sustainability. On behalf of the company she participates in the Sustainable Development Goals Action Campaign of the United Nations with the objective to win an Action Award [46]. "Our company meets international standards of Green-ness!" She gives interviews to technical magazines. The branch organization puts her in the spotlight as the best national example of how fellow companies should handle sustainability. At the same time, technical staff of the very same company continually sends out signals that they are waiting in vain for management to give them the go-ahead for starting sustainability projects. This situation exists already for years on end. There are many other companies that run R&D departments working on projects that aim at sustainability. Still, despite all of this work, multiyear development projects get shelved. Projects start and stop. New projects start and stop. Somehow these companies are unable to walk the walk. To make situation 2 understandable and manageable for you, this phenomenon requires further analysis rather than moral judgment.

In my experience, it boils down to a few fundamental aspects you need to see and understand. The first and most obvious one is economics. Put in jargon: EBITDA

(Earnings Before Interest, Tax, Depreciation and Amortization), ROI (return on investment), RTEP (real-term earning power) and NPV (net present value). These are the "Powers That Be" ruling the past, present and future of business. The modern manager checks a dashboard on his iPhone/Android to see whether these Key Performance Indicators are in the right ball park. If so, he is happy since he gets a bonus. His employees are also happy since they deliver what was agreed upon in their Annual Appraisal Form. They also get a bonus. Their families are also happy. Shareholders are also happy since they get high returns on their shares. Everybody is happy!

This is me slightly overstating my case. Still, there is a lot of truth in saying these Powers That Be are the backbone of most businesses. Don't get me wrong here. I am neither questioning nor fighting this fact. Any business that takes in raw materials and adds value by putting in human knowledge and skill to manufacture a product valuable to customers is entitled to ask more for it than just the manufacturing cost.

The core question is this: Why change a successful business? Considering the cumulative benefits for all people involved, the forces to keep doing business as usual seem to be insurmountably high. Change introduces risk. Uptime may go down. No one wants that to happen. No one wants to be responsible for a project that may lead to that outcome.

Will the set of Powers That Be remain unaltered? Of these, will ROI remain the only dimension to measure the success of a business? More specifically, will axioms like continuous growth of sales and sustained high financial returns be with us until the end of time? Looking at the internal factors discussed so far, the answer is yes. There is no reason for change. Don't change a winning team. There is simply no internal driver for a successful company to really go for innovation or sustainability. Economic growth in the present-day context is still mainstream thinking. Having said this, it is interesting to note more and more people are busy rethinking the economic model of Western countries. Explainer 2 sheds light on the growing number of models in which businesses can flourish without growing and economies can fulfill the needs of every human being on earth.

Explainer 2: Developing zero-growth economic models.
More and more economists are working on economic models redefining economics for a world in crisis. Nobel Prize winners and MIT economists Banerjee and Duflo [16] show how economics, when done right, can help us solve the thorniest social and political problems of our day. From immigration to inequality, slowing growth to accelerating climate change, we have the resources to address the challenges we face. Kelton [17] criticizes deficit thinking by showing that the question of how to pay for healthcare, creating new jobs and preventing a climate apocalypse is misguiding. She states our beliefs about deficits and the role of money and government spending are all wrong. Rather than asking the self-defeating question of how to pay for the crucial improvements our society needs, Kelton guides us to ask: which deficits actually matter?

Oxford professor Raworth [18] points to relentless financial crises, extreme inequalities in wealth and remorseless pressure on the environment. She states that anyone can see our economic system is broken. She developed the concept of Doughnut Economics. First, she identifies seven critical ways in which mainstream economics has led us astray – from selling us the myth of "rational economic man" to obsessing overgrowth at all costs. Then she offers an alternative roadmap for bringing humanity into a sweet spot that meets the needs of all within the means of the planet.

Whereas Raworth takes a universal position in her economic thinking, Schenderling [19] focuses on the Netherlands when he imagines a country that lives within the carrying capacity of the Earth in 2040. He imagines by that time the Netherlands has features like the absence of economic growth, prosperity for the vast majority of the Dutch and happier people with less stress and more free time. He touches upon aspects like changing the organization of the economy, social security, agriculture, healthcare and mobility. He translates this into practical terms for daily life: in your household, your work and your living environment. Paul introduces the concept of "Sufficiency."

Mazzucato [20] concentrates on the question of whether it is the public or the private sector that is the primary instigator of innovation. She argues that the public sector has been the boldest and most valuable risk-taker of all when compared with dynamic entrepreneurs in the private sector.

Perez [21] draws upon Schumpeter's theories of the clustering of innovations to explain why each technological revolution gives rise to a paradigm shift and a "New Economy" and how these "opportunity explosions" also lead to the recurrence of financial bubbles and crises. Examples comprise the industrial revolution, the age of steam and railways, the age of steel and electricity, the emergence of mass production and automobiles, and the current information revolution/knowledge society. She analyzes the changing relationship between finance capital and production capital during the emergence, diffusion and assimilation of new technologies throughout the global economic system. This helps in better understanding the economic problems of today.

Note: Part of this Explainer is paraphrased text originating from publishers (see quoted references).

If there are no internal factors to drive change, let us look then at external factors. Are there other boundary conditions relevant for the company coming into play? Boundary conditions in arbitrary order like limitations in physical plant plot size, a cap in CO_2 emission, the vicinity and availability of huge amounts of sustainable power, prolonged added value to the customer at large, a cap in thermal heat load of surface waters, the need to guarantee sustained long-term continuity of the business to the shareholders instead of rapid growth and returns in the short run, etc.[2]

From these questions the notion arises that external factors may affect the course of business of your company at some point in the future. This legitimates developing sustainability projects that are in the long-term interest of the company. You can contribute, thus enabling the company to start walking the walk.

Situation 3 is a peculiar one. You observe several projects in the organization that clearly pass the test of circularity. Some of them are in the R&D phase, some in the engineering phase and some in the deployment phase. Despite these develop-

---

2 Measuring economic performance in terms of growth may not be the only yardstick of business success since the Corona crisis kicked in worldwide. Scientific research has demonstrated that resilience may well become the second most important performance indicator for companies and organizations [1]. This statement is supported by the fact that several renowned consultancy companies mention resilience explicitly as one of their focus areas as is shown in Explainer 1.

ments, sustainability is not part of the value set in the company Narrative. This situation being the reverse of situation 2 also shows a discrepancy.

For about a decade, I have worked for an organization that carefully maintained the image of a fossil-based business to the outside world. In the same period, I was given all sorts of projects to manage – each of them being about sustainability! Not only that, there were several deployments of sustainable projects in manufacturing plants – but they were not called that way. They were named differently. How can this be?

One answer is that some managers inside the organization pushed their own agenda. To be successful, they made sure their project proposals were presented to higher management, respecting prevailing project evaluation criteria and the Narrative value set adhered to by the top brass. They reframed projects to get them approved. Frames that worked were safety, quality, reliability and efficiency. Or they piggybacked them onto larger projects. Apparently this can happen in a multilayered organization where middle management takes a more progressive approach than the top.

Another answer is that there was a sound understanding among the top managers about what was going to happen with the environment in the long run as a result of industrial activity. Backcasting a scenario to counteract these industrial effects, they decided to set up programs for developing and assessing sustainable technologies. They even went as far as having complete sustainable plants designed and cost estimated by renowned international engineering companies. This way, they developed roadmaps of sustainable technologies including fixed capital and return on investment, for later use, for later management, not for now. These top managers deserve praise both for their long-term view and for their courage to set up these sustainability programs well ahead of time.

How do you start if you find yourself in such a situation? I will answer that question after we have had a look at the human factor in Section 1.7.

Situation 4 has the advantage of being completely nonambiguous and fully transparent like situation 1. There is no mention of sustainability in the Narrative, no deployment of this type of project either and no discrepancy here! If you are relatively new to the company and you have an ambition toward sustainability, what can you do? This situation may feel like the fit between you and your employer is suboptimal in this respect. Again, I will answer that question in Section 1.8 after we have had a look at the human factor.

1.7 The human factor

So far we have explored the part of context which is about "hard" aspects like targets, structures and company objectives. All of these are indispensable when running a business. Yet, there is another part of company reality that people often call "soft." I call this the human factor. Interestingly, in most of the companies I consulted, people rather talk about the hard aspects than the soft ones. Next to being interesting, this observation is also quite funny since all organizations are made up of people, no company excluded. Perhaps this preference for talking about hard stuff is a typical male thing that will perish over time. Let's leave this sociological analysis to the experts.

Starting from your perspective Figure 2, the first layer of the human factor is the social network you have established. It comprises all relationships you have built with other colleagues. A lot of these will have a professional side to them. Many may also have emerged from events, gatherings and meetings that have been organized in the past. It means you have a mental map of who is where in the organization and what he or she is doing. Sometimes you do them a favor. Sometimes you get a favor in return. Next to being professional, these contacts are also enjoyable. You discover common interests, similar opinions and drive. It is fun to work with these people!

Skipping layer two for one moment, the third layer of the human factor gets closest to your skin. It comprises your value set, your convictions and beliefs: this value set you have internalized from your childhood until the present day; moral values your parents have taught you, professional values that came with your education and social values you adopted while stumbling through puberty and adolescence; trust, honesty, reliability, perseverance, love, empathy, stamina, accuracy, scrutiny and patience. Based on these values, you have views on various aspects of life in a broad sense. These may concern any subject, ranging from full-time/part-time work, to establishing a family, to traveling the world, to raising children, to road mapping your company's business, to buying or renting a house, to volunteering, to analyzing political developments in your country and to adopting religious beliefs. It is this set of convictions and beliefs that drives you, guides you through life and gives you focus.

Now comes the second layer of the human factor since it builds on numbers 1 and 3. It is the observed behavior of you and your colleagues. It also includes the bundle of idiosyncrasies that make you unique as a person, standing out from the crowd: a valuable, likable human being with wit, humor, sports practiced, hobbies enjoyed, holidays planned, etc. It is useful to realize that there is interaction between your set of convictions, beliefs and values and those of the company. It is no news stating the overlap will be less than 100%. This will be true for most employees. For you this means the actual behavior you observe of your colleagues is a compromise between these two sets of convictions, beliefs and values.

What did we learn, Palmer [22]?
- You know many people, including their preferences and ideas and drive. The professional part of these relationships is quid pro quo. On a personal level, you mutually connect which makes working together enjoyable.
- You have a personal drive and focus that is rooted in your convictions and beliefs based on a value set that has matured during your lifetime.
- Behavior you observe in yourself and your colleagues emerges from a compromise between personal and company value sets. What people communicate by means of their behavior may not exactly match what they think, admire, dream, need and want.

Your social network inside the company
Your observed behavior and idiosyncracies
Your value set, convictions, beliefs

Figure 2: The human factor.
©H.N. Akse 2025

The last point may look trivial. You may even ask: what else is new? Figure 3 sheds light on this.

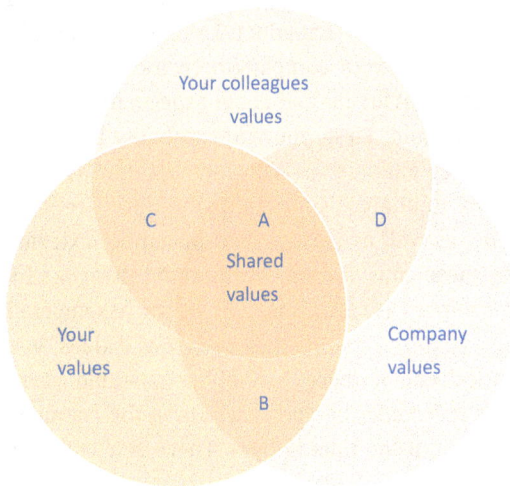

Your colleagues
values

C   A   D

Shared
values

Your   Company
values   values

B

Figure 3: Personal and company values.
©H.N. Akse 2025

Behavior based on A in the middle is what you observe on a daily basis. There may be additional values (B) you exclusively share with the company. For example, excellence may be one of your values, which matches operational excellence expressed in the company Narrative. At the same time you may have colleagues for whom excellence is not of prime importance. This makes excellence an exclusively shared value between you and the company. There may also be values (C) you exclusively share with other colleagues. Sustainability may be a commonly shared value. Values, ideas and beliefs in the latter two areas of common ground (B and C) are not likely to be expressed in the visible behavior of people. Nevertheless, their existence is possible.

My point is this: When you are in situation 3 or 4 of figure 1, you may have guessed – from observing people's behavior in the company – there may be limited potential support for your sustainability initiative. Once you start testing your idea of a sustainability project adding lasting value to the company with colleagues on various levels, you may find there is more support than you previously thought. How to test your plan comes next in Section 1.8.

1.8 Karl Friedrich Hieronymus von Münchhausen

What can you do when sustainability is not part of the company Narrative as is the case in situations 3 and 4 in Figure 1? Starting with situation 4 (no Narrative, no projects), you first have to make up your own mind about what you consider to be an ambitious yet achievable sustainability project. What that looks like I cannot possibly say, it is your call.

Next, you carefully consider which leadership style is both effective in creating change and closest to your natural modus operandi. Explainer 3 sheds light on this. Referring to Figure 3 you engage with people in your company network – Non Solus [23] – assuming what they show by their behavior may in some aspects differ from what they believe in. Your objective is twofold. First, you want to know how they think about your project. Can and will they endorse it? Will their opinion have an impact? Second, you use this opportunity to analyze your company. Here I refer to internal factors like the Powers That Be, as well as existing and upcoming external factors affecting the company. During these consultations with your colleagues together you critically examine what the company is really about. Then you ask yourself this question: will your sustainability project add lasting value? If it does, you have found a solid basis to continue. If it does not, there is simply no passable road I can advise you to take other than either reconsidering your position in the company or shelving your sustainability ambitions until the time is right. There is a saying in the southern part of the Netherlands: you can't gawk at an oven.

Figure 3 also suggests there can be exclusive common ground between your values and those of the company. Practically speaking, you may have connections with people who are the prime bearers and preservers of company values: top management. These relationships are valuable and become necessary once you have identified the right project.

Assuming you and the colleagues you consulted conclude your sustainability project does add lasting value to the company, you need endorsement by influential decision-makers and a quantified proposal. You need to present the project proposal to management. It will consist of four phases:
1. A Self-Innovation Workshop delivering the best step-up in process sustainability
2. Development of a Plant Revamp Proposal: Design to Deploy
3. Ranking of all Investment Proposals
4. Deployment when the Proposal is approved

You need budget for 1 and 2. The need to secure budget for 1 and 2 in one go I will explain in detail in Section 4.1.

What I am describing here actually is a Von Münchhausen move. Pull yourself and your proposal up until your project proposal is taken seriously. This requires mental flexibility, courage, perseverance, conviction, social and convincing skills. You are going to take personal risk, you will be out of your comfort zone for a long time, you will rock the boat, people will oppose or support you and it will feel like your job is at stake. And – as a matter of fact – it is, right from the moment you publicly announce your proposal. It is useful to realize what you are up against before you start. If you can't stand the heat, stay away from the fire. If it feels like a real challenge, take it on. Keep remembering, you are pursuing a project that will create lasting added value for the company.

Explainer 3: Types of leadership behavior that promote change and innovation.

Change in organizations has been investigated scientifically. Burnes e.a. [24] describes Lewin's three-step model of change that dates back to 1947. In Lewin's work two aspects are important when a leader wants to initiate and manage a change process in their organization. Urgency is the first. Second, to be successful in realizing the change process, the leader must show lasting and visibly different behavior. Rotmans [25] shows that 70% of successful transitions toward sustainability depend on changing our habits and behavior.

Some leadership styles are effective in promoting innovation behavior of colleagues [26]. Specifically "identity leadership" generates trust, work satisfaction and innovative behavior of coworkers [27]. This type of leadership is characterized by four elements: the leader has to be "one of us," he or she has to do everything for us, he or she has to create a "we" feeling and he or she has to put this "we" feeling into actions taken.

Another leadership style is "charismatic leadership." Quantitative research results indicate that there is a causal relationship between charisma and results. This type of leadership is effective in managing change [28].

"Autocratic leadership" has also been scientifically investigated, the main question being whether this leadership style is beneficial for economic growth. Research covered 133 countries in the period 1858–2010 [29]. It appears that autocratic leaders with positive effects on economy are found only as frequently as one would expect due to chance alone. Of course, "economy" differs from our subject of "change toward sustainability" but these findings do not add much confidence in an autocrat as a leader of change.

An international study on leadership behavior has been carried out [1] using a vast database of employee interviews owned by consultancy firm Kornferry. It covers the period of the financial crisis in 2008 and the period of the Brexit referendum in 2016. From this scientific research, emerging qualities of a good leader in uncertain times of change appear to be:
- Know how you (over)react in a crisis
- Be aware of your context, be flexible in leadership style and self-reflect
- Have your own Narrative and use this to create bonding among your people
- Deploy identity leadership: connect people, put the group first
- Organize or be the counterforce that can keep the executing powers on track
- Be modest; take leadership seriously and yourself less seriously

The key message you have to convey to key people during the Von Münchhausen phase of this process is that your project is essential for the company. The precise arguments you should use depend on what the company is really about. Being explicit about the real purpose of the company and choosing the arguments that show how your project supports the company's purpose will trigger your decision-making colleagues and higher-ups. They will grant you funding for phases 1 and 2. If the ranking exercise of all Investment Proposals in the company approves your project, phase 4 will be funded as well. You are in business!

How can I be so sure this is the way it goes? The answer is simple yet true. There is tremendous power in proposing a daring novel project that is in line with what the company is really about. People immediately feel it is good for the company; it inspires them, they want to get on board. It is because you show true entrepreneurship, which is the behavior that in all likelihood has made your company great in the past. No CEO or shareholder can oppose a project that opens windows to a bright, sustainable and profitable future. Having said this, reality forces me to add that you will need exceptional skill in maneuvering to win people for your plan. Without wanting to do any of your colleagues a disservice, you may encounter roadblocks like conservatism or jealousy hampering or blocking the progress of your project. People are people.

Finally, we have situation 3. There is no mention of sustainability in the company Narrative, but in practice there are deployed sustainability projects. The first thing I advise you to do is to talk with colleagues who initiated these projects. They can tell you how they did it. In Section 1.6, I described a setting in which sustainability projects were reframed to get them approved. These projects were sold to management using labels like safety, quality, reliability and efficiency. Although this may be a way forward, in fact it is misleading. I recommend to take the Royal Route which is in fact the Von Münchhausen move already described. This is a tougher nut to crack than reframing your project to make it sound good but your transparency will be much appreciated.

1.9 Summary

The subject of sustainability and innovation is embedded in a context you need to appreciate and understand before you can become effective in making your company's manufacturing process more sustainable.

The company's Narrative about vision, mission and identity sets the scene, including the value set everybody is supposed to comply with.

The company's structure supports business functions. It implies you need endorsement from three to ten people in different parts of the company. It also implies technology is but one element in the management toolkit. The company's structure can accidentally blind you, leaving you thinking every part of the organization is cast in stone. This blind sightedness unnecessarily may hamper your sustainability initiative.

The discrepancy factor reveals the fact there may be a gap between what a company's Narrative states about sustainability versus the existence of deployed projects. There are four possible situations. For each situation, we outline a course of action when you want to pursue sustainability projects. Here von Münchhausen rules: you have to pull yourself and your proposal up until your project proposal is taken seriously.

There are outside factors affecting the company you have to take into account. Historical examples comprise HSE standards, total quality management and compliance. More recent examples of outside factors that can affect your company are sustainability benchmarking and – in Europe – the CSRD.

The human factor is a substantial and decisive part of the context, especially when change is required. Choosing a leadership style that is close to your personality and that has proven to be effective in managing change is part of this context. Transparent strategies are outlined, which build on the human factor and optimize the chance of successfully deploying sustainability projects, depending on the starting situation you find yourself in.

2 Innovation and sustainability

Abstract: This chapter conceives a practical definition of innovation, being a new piece of equipment. It illustrates the importance of having people in your Team with adequate experience, mindset and skill sets to make innovation happen. Next, it shows the power of having the mental skill to imagine other routes to products. Subsequently, the impact of typical company phenomena like memory-fixed, memory-lost and copied behavior on your initiative is described. Then, the chapter shows how to turn a heterogeneous group of people into a highly creative Company Innovation Team. A practical definition of sustainability is chosen out of a set of three developed by the United Nations between 1987 and 2015. This chapter wraps up recommending circular instead of linear innovation.

2.1 What is innovation?

The Romans invented the word. They used the verb "innovare" which can mean two things: either the act of introducing something new or the thing itself that is introduced [30]. By inventing the verb the Romans brought something new to the table. The ancient Greeks have a word for something that eats its own tail: Ouroboros [31]. This first alinea is a nice example of an Ouroboros.

How do we define innovation two millennia later? Let's look at a few angles from which innovation has been defined. We start with a corporate perspective on innovation as described by Verloop [32] in 2004. Interesting in his book is the fact that he places innovation in a business context which turns it into a potentially powerful instrument for any company. Furthermore, he makes a clear connection between innovation and sustainability: both are oriented toward change and the future. *Harvard Business Review* in 2011 looks at innovation from an economic angle. *HBR* takes a 180° different approach by advising to drop the term innovation altogether stating that it should not be about new but about adding value. [33]. *HBR* warns its readers that innovation is not a panacea. They say, in fact, most innovations are doomed to fail. Consultancy firm McKinsey in 2022 again chooses the business angle by defining innovation as the systematic practice of developing and marketing breakthrough products and services for adoption by customers [34].

Harmsen, de Haan and Swinkels in 2018 look at process innovation from a more academic angle [35]. They consider innovation to be a three-layered concept. First it means the successful introduction of a new product or process into the market. Second, it means having management methods and systems in place to always have a consistent portfolio of innovation projects in progress. This is called innovation portfolio management. Third, it means the management of individual innovation projects into successful market introductions. This is called innovation project management. From this it is clear the authors honor their classics by adhering to the Romans' first definition: the act of introducing something new.

https://doi.org/10.1515/9783111383668-002

Interestingly, Harmsen et al. [35] adopt a stage gate approach as the backbone of their book on *Product and Process Design* in the very same manner Verloop did 14 years earlier. Their approach describes innovation as a linear process comprising discovery, concept, feasibility, development, detailed engineering and implementation. As they rightly point out, this approach is dominant throughout the process industry.

Personally, I have never grasped the logic of trying to develop something new by applying the exact same method of linear innovation for decades on end. How can we realistically expect to find something really new if we do not continually innovate the way in which we do our innovation work? In all modesty this guide therefore offers two leads for innovating the innovation process: 6-steps SIM (Chapter 3) and circular innovation (Section 2.7 elaborates on this concept). The novelty of 6-steps SIM is that it is based on bisociative thinking on an abstract level rather than starting from a newly found chemical reaction. The novelty of circular innovation is that it is connected to existing manufacturing routes and it recurs continually, whereas a stage gate approach assumes innovation happens only at the very beginning.

What do we mean by innovation in this book? We stick to the second definition of the Romans: the thing itself is new. Why do we choose this definition of innovation? First, this is caused by your situation. You are dealing with an existing manufacturing process that has been up and running for 20–60 years. Your process needs to become more sustainable. The average time to complete a stage gate approach from discovery to deployment is 10 years in an optimistic scenario and 15–20 years in a more realistic scenario. Even if your R&D colleagues have already made a promising discovery 5 years ago that can enhance the sustainability of your process, you do not have that sort of time on your hands when it comes to meeting a zero-emission CO_2 target in 2050.

Second, describing innovation as a new thing makes it tangible. Everybody can see it, touch it, buy it and deploy it. It can become a new entity in your plant that was not there before. Does this mean innovation is short-circuiting: simply buying new stuff to replace old stuff? Not quite! Chapter 3 describes and explains the Self-Innovation Method that guides you through various stages of the innovation process. For now, I like to share some other aspects of innovation with you that are important when you want to become successful.

2.2 Company maturity and required mindset and skill sets

Most companies start as a small business and tend to grow over time. This is not only true for well-known established companies. It is also true for companies that are in start-up or scale-up mode. One may question the underlying axiom why continuous growth is necessary for all businesses. I refer to Explainer 2 for answers to that question. Here I like to make another point. It is this:

When you run a company in start-up/scale-up mode, you need people with a specific mindset and skill sets. To name a few, one must have the following qualities in arbitrary order: flexibility, resilience, perseverance, open for opportunities, creativity, a "do–it-yourself" attitude and skills, the capacity to improvise, the need to cooperate, a focus on customer needs and the ability to translate this into manufacturing requirements.

When you are part of an established company with an existing customer-base your company needs people with different mindsets and skill sets. Control over operation, management of risk, strict adherence to safety regulations, compliance with procedures, attention for reproducibility of performance, preservation and timely maintenance of assets and more.

To illustrate my point, please have a look at Figure 4.

Figure 4: Masjid-i-Suleiman and Geleen.
(a) © BP [36], (b) © IA Professionals [37]

On the left you see a BP oil derrick in Masjid-i-Suleiman (Persia) in the early morning of May 26, 1908. The whole camp suddenly started to reek of sulfur. At 4 o'clock the drill reached 1,180 ft and a fountain of oil spewed out into the dawn sky: First Oil! On the right you see Sabic's Olefin-3 cracker in the southern part of the Netherlands.

The BP people had to act fast, think on their feet and come up with first-time solutions to get this rig under control in a situation where virtually nothing was available. The Sabic people maintain a well-established workflow in keeping a five–decade-old naphtha cracker up and running. In each plant control room on-site the daily routine is a Chinese copy of the day before. All efforts are focused on meeting olefin production targets by adapting process conditions based on varying naphtha composition. Obviously, these two situations are extremes. As we now know, BP's oil rig was the onset of a new and hugely important business – at that point still in its infancy. The Sabic cracker is a nice example of a business and a technology that have become fully mature.

Why is this concept of maturity important to you? Like I assumed earlier, your manufacturing process will be between 20 and 60 years of age. This implies that business and process are mature. You will recognize recurring routines, fixed procedures, a jointly supported idea of how you and your colleagues arrange things, how you start and manage projects, how you evaluate ideas, etc. Within the actual context these routines have proven to be the most effective ones in order to meet business targets.

Having a leading role, you now come up with the proposal to start an innovative project to increase the sustainability of your manufacturing process. Preliminary desk studies carried out by your Team indicate that it will surely lead to changes in process hardware.

Clearly, your initiative is not part of the daily routine. Also – mindset-wise – the people you need to turn this proposal into a success should look more like the oil explorers in Masjid-i-Suleiman in 1908. Open and creative, entrepreneurial, ready to come up with solutions for unexpected situations and eager to succeed.

At the same time, you need colleagues in your Team that have in-depth knowledge of the process part your proposal is focusing on. Usually, they have been working in the plant for years. They know the intricacies of plant behavior, how to handle them, the limitations and peculiarities of the process. I am talking about (chief-) operators, technologists, (assistant) plant managers, maintenance people and lean six sigma Black Belt people.[3]

The Self-Innovation Method anticipates this duality. On the one hand you need novelty-minded colleagues with the ambition to innovate. On the other hand you need experienced colleagues that can operate the plant blindfolded, knowing it front to back. This Operational Experience is the first pillar of the Self-Innovation Method. My observation has invariably been that asking an experienced Team the right sequence of questions opens up a totally new mental window of opportunities. In this window, creative and experience-based types of thinking mutually stimulate each other to come up with truly innovative and deployable solutions. In Chapter 3 you will find these questions.

2.3 Mind over matter

Some philosophy has to be part of this guide. Weren't you impressed – even intimidated – when you had your first tour through a plant? The sheer size of process units, the complexity, the dazzling sound of steam, the faint smell of chemicals, the immense heat radiated by hot sections, the sudden flares of burning excess hydrocarbons and the seemingly endless pipe bridges connecting the plant to feedstock and storage tanks at the harbor. I surely was intimidated when I got my first tour on a chemical site. Seeing these huge structures made from concrete and steel covered by shiny aluminum plating almost automatically triggered the notion it would last forever, fixed, cast in stone.

---

3 When coaching Teams I often add up the years of working experience of all participants. With groups of about 10 people you easily have something between 150 and 180 years of experience with your manufacturing process in one room, which is something of unique and great value.

Following logic, however, we know that there has been a time when these plants were not around. And then – interestingly – there was a time they only existed in peoples' minds. Isn't that fascinating? Mind comes first, then matter. This sometimes is hard to grasp when your daily environment is made up of existing kit.

It means that there is a lot of power in having the mental skill to imagine other routes to products. I like to call this Imagining Skill. It is the second pillar of the Self-Innovation Method. It comes right after the first pillar of Operational Experience mentioned in Section 2.2. Be aware the mental exercise of creation and innovation always precedes design and construction of a plant. New plants don't drop from the sky. This pillar of imagining skill makes SIM a powerful tool.

The third pillar of the Self-Innovation Method is Working Knowledge of Physics and Chemistry. What do I mean by "Working Knowledge"? It is the accumulation of insights of how fundamental physical and chemical phenomena (like transfer of heat, mass, impulse and occurrence of reactions) work out in the daily practical environment of a chemical plant. Working Knowledge has a hybrid nature.

It combines theory of natural sciences you have learned during your education with hands-on practical know-how you have learned in the plant. It also has a self-cleaning nature. Nice theories that cannot be deployed in the plant are quickly removed from engineering consciousness. On the other hand, practical procedures that work in the plant lacking a sound theoretical basis are maintained: I know it works, but I don't know how it works. The best jobs to build Working Knowledge are at the interface of technology and plant operation. Once matured, Working Knowledge supplies its owner with a sense of direction, a gutfeel of what might work and what will never work. This is of great value to the company!

The fourth pillar of the Self-Innovation Method is Technology Overview. It comprises all known process technologies including the latest mature ones that are still scarcely deployed. This overview not only serves the purpose of giving quantitative information you can base your calculations on. It also serves as a source of inspiration. Appendix 2 and Table 14 give an example.

Mind over matter also holds for the immense number and variety of simulation and calculation tools that have become available for physicists and chemical engineers in the past decades [38]. The temptation to install a computer simulation program on your system, throw in some process data and run it to see what comes out is almost irresistible. It's so fast! Computational fluid dynamics (CFD) looks so nice! (Figure 5). Product purity exiting distillation columns can be calculated up to five decimals behind the comma!

We take a bold position here. Innovation does not come from tools, however fast and beautiful they may be. There is one exception, being the category of programs that can produce lots of different Process Flow Diagrams in a short time span based on reliable physical and chemical properties [39]. Although these programs can generate unexpected and internally quasi-contradictory PFDs, it still is the experienced chemical engineer who has to make the final call which line-up is the best. As for the rest, tools can be very effective in speeding up each process design phase. The point is that they come in after the innovation has been generated. Even then, software tools need to be treated with caution, a healthy dose of skepticism and common sense. Otherwise, CFD may easily turn into Colors for Directors [40]. In short: Innovation comes from people.

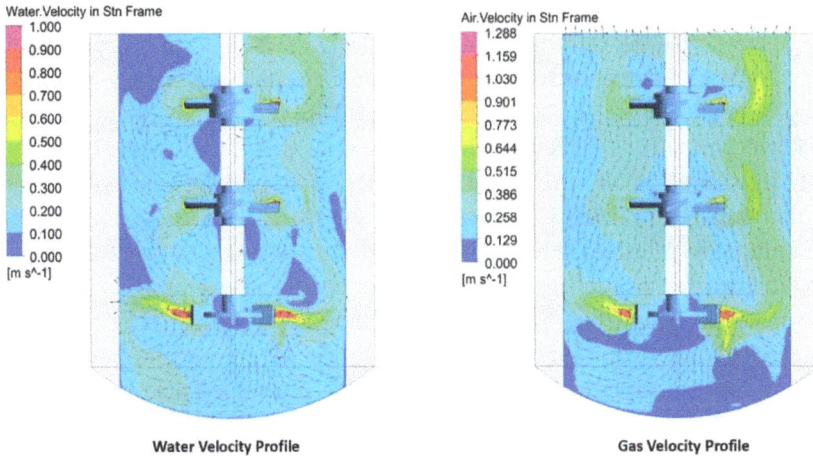

Water Velocity Profile   **Gas Velocity Profile**

Figure 5: CFD output shows water and gas velocity profiles. But who got the idea of calculating these profiles in the first place?

In summary, the Self-Innovation Method is built on four pillars:
- Operational Experience
- Imagining Skill
- Working Knowledge of Physics and Chemistry
- Technology Overview

By this time the question may have crossed your mind what these pillars mean for your innovation project. To be more specific, what does this mean for your Team composition? Section 3.2.3 deals with the Company Innovation Team in terms of diversity and job types of Team members. Here I like to answer this question by looking at the four pillars quoted. In fact this is straightforward. You will need people with hands-on operational experience in the plant. Next, you will need people who are great in imagining skill. Third, you need people with proven Working Knowledge. Finally, you need people with a broad technology overview ranging from well-established to state-of-art novel technologies (Appendix 2).

It goes without saying that each of your colleagues will possess combinations of these different knowledge set and skill sets. For you, this means, you have to find out upfront which of your fellow workers excels in which areas. Only then you can put together a

Team that is competent to take on the challenge of innovating part of the manufacturing process.

Having said this, once in a while one comes across a white raven that can do it all: A Walking One-Man Innovation and Deployment Department. I got to know such a man while I was in a job interview with the CEO of a midsized family-owned biotech company. The ice had just broken and we were in the midst of a lively discussion about the question whether the future of manufacturing technology would be shaped by trendy enzymes or old school metal catalysts. Suddenly the door crashed open with a loud bang. A man burst into the room dressed in a blue sweater and jeans, walking on clogs. Impatient and agitated he stomped to his CEO with the announcement that the construction department was building a device against his wishes. Our conversation immediately fell silent. The CEO exchanged a few words with him. He then agreed that construction would immediately be halted until further notice. Satisfied with the result, the man stomped back out of the room. As I learned later, he managed the R&D department, the construction department, the engineering department and – this is true – the marketing and sales department. Naturally, each of these departments was led by a dedicated manager. This Organisationsstruktur however did not prevent him from calling the shots. Anytime, anyplace, anywhere. Never have I seen novel technologies being developed from TRL = 1 to TRL = 9, marketed and sold in large numbers so fast because of his interference. Despite these results, I'm not sure I want to recommend this as the best way to innovate. And despite this watershed in the interview I got the job.

2.4 Company memory and copied behavior

Maturity not only impacts the required mindset and skill sets of the people to run the company effectively. In the process of becoming a mature organization a phenomenon called "company memory" also develops. Company memory contains a whole array of meaningful events. These can be things that have positively impacted the company like acquiring the first large business account, the first big plant extension, experiencing years of sustained autonomous growth and running a profitable business. Inevitably negative things will also be a part of company memory like incidents in plant operation, economic bad times, unexpected shut downs and energy crises. Everybody in the company knows these things. They are part of the company's pride and they are part of the tough lessons learned.

One specific part of company memory relevant for you comprises all memories fixed. They can lead to roadblocks on your way. Often these concern technological plant changes that have been proposed in the past. Like you and your coworkers, earlier generations of technologists have been trying to improve the process. All their successes and failures are stored in company memory for eternity. You will surely encounter these when you publish your sustainability initiative. Company memory will generate endless series of arguments why your initiative is doomed to fail. "This has been tried by mr. Johnson in 1986 and it did not work." The best thing you can do is to check the proposal and the context of Mr. Johnson in 1986. As it often turns out, proposals are not really identical and situations are not really comparable. Still, you can learn from what your predecessors have done in the past to improve your proposal in the present.

There is also a complementary part of "company memory" which is equally interesting and impactful. It is the collection of memories lost. Let me give you an example from consultancy practice. One day, we were consulting a large chemical company in the Rotterdam area in the Netherlands. After defining the scope of work with company employees, my partner and I checked the Process Flow Diagram for opportunities to enhance process sustainability. At some point we came across a large vessel. It was a few meters in diameter and over four meters in height, made of 316 stainless steel. A large pressurized gas stream entered the vessel perpendicular to the side to leave it axially over the top. This 90° turn in gas flow direction caused a large pressure drop. For this reason a huge gas compressor had been installed to keep process pressure at the right level. But now the amazing thing comes. This vessel was totally empty! There were no internals, no catalysts and no nothing. It was a significant piece of equipment in terms of size, material type, concrete foundation and ducts. It had cost a significant amount of money to build. It still generated significant operational cost in the form of electricity consumption and maintenance of the installed compressor. We therefore asked our company counterparts what the function of the vessel was. After a deafening silence the answer came. The vessel had always been part of the plant and there was no reason to remove it. Apparently, company memory can retain certain memories forever and blank out others completely.

There is a third aspect of company memory you have to deal with. Let's call this copied behavior. It is difficult to detect when you are part of the workforce. We tend to do things the way we are told; they have always been done without questioning or revisiting the reason behind it, even long after that reason has ceased to exist (Explainer 4, cited ref. [41]). The assumption behind this example is that human behavior will resemble the nonhuman behavior described). The existence of this phenomenon is relevant for you to know. It means you can question aspects like working procedures that are normally not questioned since it is an unwritten law in your company not to question these things. Normally only outsiders – like us, consultants – are allowed to ask these questions. My advice is "don't worry about this. Pop the question!"

What is my message to you when describing memory-fixed, memory-lost and copied behavior? The feature these aspects have in common is that they will affect your innovation initiative. This may take on different shapes. It may hamper you; it may blur your vision; it may create an atmosphere of doubt. Make no mistake. Each and every company has these memory and behavior examples, even start-ups and scale-ups. This is all the more relevant since memories and behavior are deeply rooted in the company's DNA. They are carried out by almost every employee. Therefore they are very strong and convincing. And they will not go away.

I want to make you aware. Be prepared and have these matters on your radar so you can respectfully and consciously deal with them as you progress with your innovation project.

> **Explainer 4: Five monkeys in a cage [41].**
> **Five monkeys were placed in a cage as part of an experiment.**
> In the middle of the cage was a ladder with bananas on the top rung. Every time a monkey tried to climb the ladder, the experimenter sprayed all of the monkeys with icy water. Eventually, each time a monkey started to climb the ladder, the other ones pulled him off and beat him up so they could avoid the icy spray. Soon, no monkey dared go up the ladder.
>
> **The experimenter then substituted one of the monkeys in the cage with a new monkey.**
> The first thing the new monkey did was try to climb the ladder to reach the bananas. After several beatings, the new monkey learned the social norm. He never knew "why" the other monkeys wouldn't let him go for the bananas because he had never been sprayed with ice water, but he quickly learned that this behavior would not be tolerated by the other monkeys.
>
> **One by one, each of the monkeys in the cage was substituted for a new monkey until none of the original group remained.**
> Every time a new monkey went up the ladder, the rest of the group pulled him off, even those who had never been sprayed with the icy water.
>
> **By the end of the experiment, the five monkeys in the cage had learned to follow the rule (don't go for the bananas), without any of them knowing the reason why (we'll all get sprayed by icy water).**
> If we could have asked the monkeys for their rationale behind not letting their cage mates climb the ladder, their answer would probably be: "I don't know, that's just how it's always been done."
>
> **Note 1:** The authors of Intersol in Ottawa suggest that human behavior in companies resembles the monkey behavior described in this case. In doing so they do not intend to be disrespectful, neither to monkeys nor to humans in the organizations nor to you as a reader. So please do not feel offended by this case.
> **Note 2.** Some sources are adamant in stating that this experiment has never taken place.

2.5 Sparking dynamic interaction in the group

As pointed out in Section 2.3, the Self-Innovation Method preferably is carried out by a heterogeneous group of people in your company. Operators, technologists, maintenance people, (assistant-) plant managers, colleagues from R&D, lean six sigma Black Belts. Usually, these people know each other well and oftentimes for years. This is both a positive and a challenging thing. The positive side is that they already have developed a working relationship that speeds up mutual introduction. The challenging side is that everyone has a mental picture of the observed behavior of others and everyone tends to show the behavior others are used to. This brings clarity; it creates safety and keeps everybody in their comfort zone. It implies that everyone shows certain personal qualities to the group and keeps other qualities behind the curtain. It also brings an implicit rule what people can say and what they cannot say. This rule is often connected to the functional role someone has in the company. A technologist is not supposed to make a statement about the strategy of the Marketing Director. A sales manager is not supposed to propose a process design change in the plant. In summary, all relations between group members are clear, safe, comfortable, predictable and therefore effective for the company. They are also static in time. Interpersonal dependencies and mutual expectations are fixed. It is no surprise to say that it will be difficult for this group to innovate.

Over the years I have learned and seen the following. To make a group of people innovative you first need to stimulate and create dynamic interaction and enthusiasm. The moment people suddenly discover new and unexpected qualities of colleagues, an enormous amount of positive energy is generated. Everyone is amazed and enthusiastic about the qualities of well-known colleagues that have always been kept behind the curtain. I have seen colleagues finding out that they had been traveling to the same spot they loved in the same foreign country for years on end without knowing it about each other. Others discovered that they were born in the same small town in the countryside and went to the same school, one 5 years after the other, knowing the same neighbors in the same street. People finding out they loved the same sport or played the same musical instrument. Everybody discovered new sides of colleagues they already knew for years! The energy this produces fuels the innovation process which is exactly what you need. It loosens existing connections between colleagues and paves the way for people to look at each other with fresh eyes.

How do you spark this dynamic interaction in your group? After much trial and error I can recommend three tools that invariably have a desired and lasting impact. My message is "First things first". To be effective, you need the right atmosphere for innovation work. Start with the tools described below.

2.5.1 Group Photo

The first tool you can use is Group Photo [42]. If you take outside courses occasionally to master new know-how or skills, you will have noticed many training centers taking pictures of the group you are part of. A few weeks after the course an envelope drops on your doormat. You open it and participants stare you in the face. You even remember the names of some of them. That was a nice course!

There are more effective ways to use photography. Before the Innovation Session starts, you install a photo booth with tripod, camera and wireless flashes in a separate room. Make sure you use a wide angle lens. You then ask all participants to distribute themselves in a row perpendicular to the viewing angle of the camera: taller people in the back and smaller people in the front. You tell them that you will make a series of trial photos to check the lighting and to make sure everyone is visible from head to toe. It is important for all participants to look at the camera and to say "cheese" at the right moment.

It is interesting to see how everybody arranges themselves (Figures 6, 7 and 8). Without any instruction, everyone keeps the same distance of 30–40 cm to their immediate neighbor. The unwritten Law of Personal Space. You ask for attention, give the "cheese" signal and take pictures. Out of ten, one or two are usable. These are the ones on which everybody has his/her eyes open and looks at the camera. You then move the camera closer to the group and ask if the outer people can move to the inside because you want everybody on the picture. With some hesitation everyone accommodates. This sequence of moving forward and taking pictures you repeat a few times. Already at the third time laughter emerges from the group. People start looking at each other instead of the camera. A sense of awkwardness is in the air. Some ask the photographer if this is really necessary. Yes, it is really necessary. A few times more, and the photographer is really close to the group. The participants are even closer to each other. The photographer says the quality of the pictures is improving rapidly. No one really hears this since everybody is now laughing loudly and talking about this uncomfortable situation of lost personal space with shoulders touching.

Figure 6: **Group Photo**: a pixelated picture taken during a SIM Workshop with a large size chemical company which explains the Team size of no less than 15 people. Team members have visibly crossed each other's personal space. Spirits are up. Yet, some people still have a closed posture. Group Photo creates the right atmosphere when given sufficient time.
©H.N. Akse

The session takes about 15 min. Its effect on group spirit and interaction lasts for 2–3 days. This is ample time to carry out the Self-Innovation Workshop with your Team. Again, I know it works but I do not know how it works. My guess is, everybody experiences the same event and shares the same feelings – clearly outside their comfort zone. This causes individual Team members to merge into a group. Rest assured, this always works!

Figure 7: **Group Photo**: a pixelated picture taken during a SIM workshop with a mid-size chemical company: a Team of six people. Body language tells you two gentlemen are not yet at ease. Group Photo creates the right atmosphere when given sufficient time.
©H.N. Akse

Figure 8: **Group Photo**: a pixelated picture taken during a SIM workshop with a mid-size environmental company. Of six people, one person still stands out from the crowd. Group Photo creates the right atmosphere when given sufficient time.
©H.N. Akse

2.5.2 Role Play

© Timon-Studler – unsplash.

A second proven tool is Role Play. You ask the group to form a circle. You instruct the first person to think about a situation and a role in which he or she is going to speak to his/her neighbor. Upfront, nobody knows the situation or the role. The person spoken to has to guess what the situation is, immediately take on his/her own role and start giving answers from this role. After a minute, the next sequence starts. The person spoken to turns 180°, thinks up a situation and a role and starts to speak to his/her neighbor. You repeat this until everybody has gone through the sequence. I have witnessed an employee talking sternly to his manager, saying: "This is the second time, I got you driving at 60 miles/hr this week and you know it is 30 max. You got away with a warning on Monday, but now I am going to fine you. You are lucky I am not busting you or confiscating your driver's license!" The manager showed rarely seen timid behavior. The next guy asked the third guy: "How do I get to London from here on my bike?" This exercise stimulates creativity and it promotes thinking on your feet. It can offer people the opportunity to take on a refreshing hierarchical role, demonstrating and releasing possible tensions in the group. If you know about hierarchical issues that hamper effectiveness, Role Play may be a tool to make this visible and discussable. At the same time it can be stressful for those among us who are less fast. Some people simply freeze when given the speaker role. You should therefore carefully consider whether this tool matches the people in your group. You should be prepared to mitigate unwanted side effects like somebody dropping out because of an apparent inability to do this in a relaxed and playful manner.

2.5.3 Map

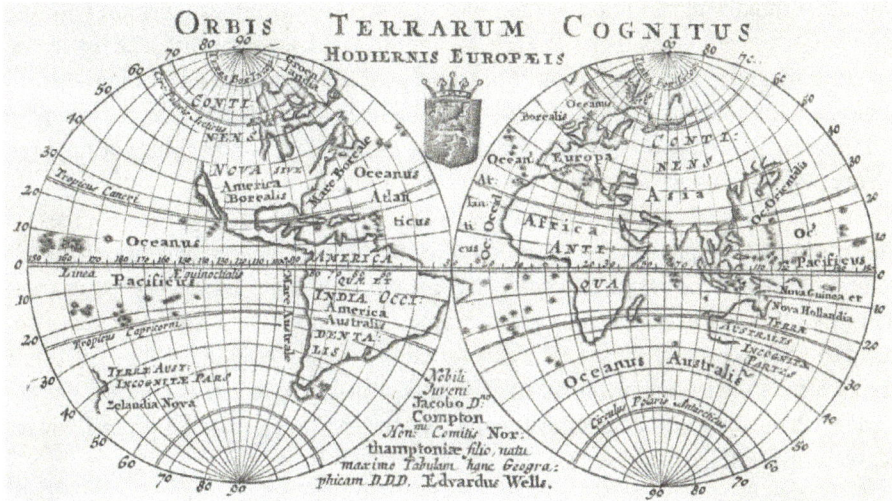

© The-New-York-Public-Library – unsplash.

A third proven tool is Map. This requires a room with quite some floor space (typically seven by seven meters), a lot of flip over sheets, a set of markers and lots of duct tape. First you fix sufficient flip over sheets on the ground to cover 25 m² of the floor, hence the duct tape. Then you appoint two Geographic Masterminds. These people you ask to draw pieces of country maps Team Members originate from. Then, everybody gets a differently colored marker and may indicate important places on the map where he or she is born, went to school, did their masters and/or PhD, had their first job, met their current partner (if appropriate) and works now. Finally, you ask everybody to put one foot on their place of birth and the other on another place important in their life.

The result is a living statue of people entangled in a seemingly impossible, hard to sustain and rather embarrassing configuration. You can be sure this sparks fun and enthusiasm, 100% guaranteed. Then, one by one, people get the chance to explain why their feet are on the spots they have chosen. Everybody can tell his or her personal story. The statue is kept as it is until all stories are told.

This tool, like Group Photo and Role Play, brings people outside their comfort zone, outside the normal course of business. One effect is people become alert and open to this new situation. Another effect is people learn a lot of new facts about each other. So, although being outside the comfort zone, everybody gets the new experience and information directly from the horse's mouth at the same place and moment in time. This shared experience creates a new type of bonding which was not there before. It enhances the drive to achieve results as a group.

2.6 Trust and safety

One other important condition has to be met before your group can kick off with the innovation process. This is the explicit mentioning and safeguarding of trust in the group. During the SIM Innovation Workshop people are invited and encouraged to come up with new ideas. These can be breathtakingly smart. Sometimes these can be unintentionally beside the point. Eternal dismay and embarrassment lure around the corner. Reputations are at stake.

You can create trust the following way. First, you make sure everyone agrees on the rule "What happens in the room, stays in the room." Only the agreed outcome is communicated to the rest of the company. Second, you make sure everybody agrees with the rule, in a brainstorming phase stupid ideas do not exist and people are advised to keep positively meant criticism for later stages of scrutiny. These rules also create a safe environment for all participants. Third, you lead by example. For instance, when someone outside the Team asks you how the innovation project advances, you respectfully stay silent.

2.7 Circular innovation and Business Evaluation Cycle

These are the four phases of your project proposal (Section 1.8):
– A Self-Innovation Workshop delivering the best step up in process sustainability
– Development of a Plant Revamp Proposal
– Ranking of all Investment Proposals
– Deployment when the Proposal is approved

If you are successful in each phase the final deliverable is deployment and a resulting increase of sustainability of your manufacturing process. The interesting question is how will this develop further after deployment has taken place? I can identify two options:
– Business as usual. The product is marketed and sold like before albeit with a higher degree of sustainability. Your project has been a one-off.
– Recurring innovation. Your company repeats the project every few years to innovate other parts of the plant.

Drivers for recurring innovation can come from inside or outside the organization. Off course, I have no knowledge of drivers inside your company other than your person. Let's therefore zoom out for a moment and look at the whole of the company's business. It may be that your company – like many others – instigates a Business Evaluation every so many years. During this evaluation all ins and outs of the business are screened. Strengths and weaknesses of the products are checked as well as market opportunities and threats. The whole product portfolio is analyzed. Margins per product type are calculated. I have seen clients who to their dismay discovered

80% of the profit was made with 20% of the products manufactured. The outcome of such a Business Evaluation can be of a commercial nature. In the given example, the company decided to drastically restructure its product portfolio. Most of the loss-making products were discontinued. The outcome can also be of a technical nature. Product quality has to improve or specifications have to be more stringent. The outcome can also point in the direction of sustainability. Customers may demand a verifiable higher degree of sustainability of your products.

If recurring Business Evaluation is procedure in your company, it is a perfect opportunity to extend it with a recurring Process Evaluation procedure. I mean you can team up with your business colleagues in a continual joint effort to improve both product and process aspects of the business every 2–3 years. This is what I meant in Section 2.1 with cyclic innovation. Having a Business Evaluation procedure in place lowers the threshold of expanding it with a Process Evaluation focusing on innovation and sustainability aspects.

Should a Business Evaluation routine not exist, then there is always the possibility of introducing one. However, for you this may be a step too far. Before this happens my guess is that there will be other drivers for recurring innovation. These come from outside. An important one is the Solomon benchmark [5] which we touched upon in Section 1.5. Such a benchmark tells you how your plant performs compared to the competition. Companies like Solomon were among the first to use measured data to improve efficiency, reliability and profitability in the energy industries. Traditionally these companies focused on expense/margin optimization. More recently, Solomon has expanded its scope to sustainability topics. Table 2 shows the selection of Solomon Sustainability Benchmark.

Table 2: A selection of Solomon Sustainability Benchmark topics [7].

1. Sustainability investment analysis
2. 2023 Olefin study to feature enhancements with a focus on sustainability footprint
3. Sustainability performance analysis
4. Carbon emissions and sustainability consulting
5. Sustainability metrics added to the liquid pipelines and terminals studies
6. 2022 Power study to feature enhancements with a focus on sustainability
7. Portfolio sustainability analysis
8. Upstream operator improves energy efficiency and GHG management performance gaps with benefits estimated at 48M USD/year
9. Oil and gas carbon emissions dashboard
10. Updated aromatics study launches with new sustainability metrics and deeper insight to help you reduce carbon emissions and deliver on your climate commitments
11. Sustainability strategy insight
12. Energy transition insight
13. Sustainability
14. Overcoming obstacles to improve energy efficiency in refining
15. Sustainability performance analysis product sheet

The fact that this Sustainability Benchmark exists means that there is a peer group of companies out there spending time and money to improve its performance in sustainability terms. Imagine a situation in which your company is not part of this group. It means the competition gets better in sustainable manufacturing each year whereas your company stands still. Examples of subjects analyzed by benchmarking companies are summarized in Table 3 [4–7].

Table 3: Sustainability Benchmark – climate-related-sustainability factors.

Energy balances
Emissions data
Hydrogen systems
Electrical optimization
Technology options
Feedstock selection
Products and logistics

In my view, this outside development is a good example of an external factor calling for a continual recurring innovation effort pursuing process improvement. In summary, I can state this. Either you can embark on an already-existing Business Evaluation Cycle by extending its scope to Process Evaluation or you anticipate increasing outside pressure to instigate recurring innovation. In both situations you will have to convince the higher management to make Process Evaluation and Innovation cyclic. I hope to have given you a set of sound arguments to make your case.

2.8 What is sustainability?

When you start to work on an innovation and sustainability project it makes sense to have a commonly shared idea about what exactly sustainability comprises. Looking at the multitude of definitions in the public domain I have decided to limit myself to those given by the United Nations since they kicked it off. More precisely, the World Commission on Environment and Development (WCED) of the United Nations was the first in giving a definition of Sustainable Development in October 1987. It can be found in the report "Our Common Future," written under supervision of Mrs. Gro Harlem Brundtland [43].

[1987]
Sustainable development is development that meets the needs of the present without compromising the ability of future generations to meet their own needs. It contains

within it two key concepts. Remarkably, most of the time these key concepts are left out when this definition is quoted in literature:

a) the concept of "needs", in particular the essential needs of the world's poor, to which overriding priority should be given

b) the idea of "limitations" imposed by the state of technology and social organization on the environment's ability to meet present and future needs.

This definition explicitly points at two key concepts: needs and limitations. I fully understand, respect and subscribe Mrs. Brundtland's emphasis on poverty in her role as Chair of the WCED. Having said this I take the liberty of interpreting "needs" in such a way that it makes sense to you. Here it is.

"Does your manufacturing process deliver a product for which there is a real need at the end of the value chain in a context given in Explainer 5?" The answer has to be given by you before the onset of your project.

The other concept – limitations – is equally important. Brundtland says that the state of art of our technologies and the way in which we organize ourselves socially impose limitations on the ability of the environment to meet present and future needs. This is a remarkable definition since you would expect things to be the other way around. We are inclined to say it is the environment that imposes limitations on the world community in its pursuit of work, food, housing, education, etc.

Assuming Mrs. Brundtland's definition of limitations is correct and translating this to your situation we can draw two conclusions:

– There are and always will be limitations (ecological footprint, resource use, energy use) your company has to respect when meeting current and future needs of the final customer.

– By changing your company's technology you can remove a certain limitation until you encounter the next one. This opens the door for cyclic innovation.

I consider these two points to be useful for you as starting points for your sustainability project.

The concept of sustainability is a living entity. This means new definitions keep popping up. I briefly touch upon two other UN-definitions published since 1987 in my quest to find a useful definition of sustainability for you.

[1992]

"Improving the quality of human life while living within the carrying capacity of the supporting ecosystems" [44]. This high-level definition has the charm of brevity. That's it. I am not able to translate this into anything you can use in your sustainability initiative.

[2015]

Figure 9: Sustainable Development Goals [45].

The SDG concept (figure 9) is hugely popular in industry. There is an SDG Action Campaign. Companies and institutes can win Action Awards [46]. The SDG website clearly shows that this UN initiative mobilizes large groups of people who actively work on various SDG topics. This in itself is a great achievement beyond doubt. It is exactly what we need: people actively take responsibility for sustainability in all of its appearances. The UN deserves tribute for this achievement!

However, when we look beyond the program's great quality of sparking worldwide enthusiasm to find the content you can use in your innovation and sustainability project, the ice gets pretty thin pretty soon (Explainer 6).

Analyzing the UN website for content of SDG 9 with the promising title "Industry, Innovation and Infrastructure" [47] the quote suggests that industry at large is a panacea that will make all SDG dreams come true. Innovation is mentioned only twice in a text that breathes the atmosphere of a group hug. Clearly, this is a political text lacking the sharpness of analysis in the Brundtland report written almost three decades earlier. Should you want to know more about SDG, other authors have written extensive reviews of the SDG concept from different angles [48].

SUSTAINABLE DEVELOPMENT GOALS
ACTION
CAMPAIGN

Figure 10: United Nations logo of its action campaign promoting Sustainable Development Goals [46].

My honest and respectful commentary on this popular UN SDG initiative is this. It works wonders when you want to mobilize, inspire and connect people all over the world to actively start thinking about and working on sustainability. Also, if you want to win high-profile Sustainability Action Awards (Figure 10) with your company this is definitely the way to go. However, do not expect miracles when you are looking for SMART guidelines that help you in setting up and deploying your sustainability initiative inside the company. My fear is that the content part of this initiative may well end up as no. 101 in the list of fads and fashions quoted in Table 1. The fact that only one out of seven top tier consultancy companies use the SDG concept to build business cases (Explainer 1) may be a heads-up for you.

In summary, looking for definitions useful for you, I recommend to use Mrs. Brundtland's description. I think her team has given the best definition since it includes needs and limitations. It boils down to this:

Does your manufacturing process deliver a product for which there is a real need at the end of the value chain?

There are always limitations like ecological footprint, resource use and energy use your company has to respect when meeting current and future needs of the final customer.

By changing your company's technology you can remove a certain limitation until you encounter the next one. This calls for cyclic innovation.

Explainer 5: Quotes ex Brundtland Report (1987) describing the context of sustainability [43].
(page 16): Humanity has the ability to make development sustainable to ensure that it meets the needs of the present without compromising the ability of future generations to meet their own needs. The concept of sustainable development does imply limits – not absolute limits but limitations imposed by the present state of technology and social organization on environmental resources and by the ability of the biosphere to absorb the effects of human activities. But technology and social organization can be both managed and improved to make way for a new era of economic growth.

The Commission believes that widespread poverty is no longer inevitable. Poverty is not only an evil in itself, but sustainable development requires meeting the basic needs of all and extending to all the opportunity to fulfil their aspirations for a better life. *A world in which poverty is endemic will always be prone to ecological and other catastrophes.*

(page 39:) Sustainable development seeks to meet the needs and aspirations of the present without compromising the ability to meet those of the future. Far from requiring the cessation of economic growth, it recognizes that the problems of poverty and underdevelopment cannot be solved unless we have a new era of growth in which developing countries play a large role and reap large benefits.

(page 41): Even the narrow notion of physical sustainability implies a concern for social equity between generations, a concern that must logically be extended to equity within each generation

(page 42): Living standards that go beyond the basic minimum are sustainable only if consumption standards everywhere have regard for long-term sustainability. Yet many of us live beyond the world's ecological means, for instance, in our patterns of energy use. Perceived needs are socially and culturally determined, and sustainable development requires the promotion of values that encourage consumption standards that are within the bounds of the ecological possible and to which all can reasonably aspire.

(page 49): Yet it is not enough to broaden the range of economic variables taken into account.

Sustainability requires views of human needs and well-being that incorporate such non-economic variables as education and health enjoyed for their own sake, clean air and water, and the protection of natural beauty. It must also work to remove disabilities from disadvantaged groups, many of whom live in ecologically vulnerable areas such as many tribal groups in forests, desert nomads, groups in remote hill areas, and indigenous peoples of the Americas and Australasia.

(page 49): Changing the quality of growth requires changing our approach to development efforts to take account of all of their effects. For instance, a hydropower project should not be seen merely as a way of producing more electricity; its effects upon the local environment and the livelihood of the local community must be included in any balance sheets. Thus the abandonment of a hydro project because it will disturb a rare ecological system could be a measure of progress, not a setback to development. Nevertheless, in some cases, sustainability considerations will involve a rejection of activities that are financially attractive in the short run.

Note: Explainer 5 is long. Yet I deem it necessary to quote the pages mentioned in full to give you a full appreciation of the straightforwardness of the text.

Explainer 6: SDG 9 (2015) Industry, Innovation and Infrastructure [47].

Inclusive and sustainable industrial development has been incorporated, together with resilient infrastructure and innovation, as Sustainable Development Goal 9 in the 2030 Agenda for Sustainable Development.

Both the 2030 Agenda and the Addis Ababa Action Agenda focus on the relevance of inclusive and sustainable industrial development as the basis for sustainable economic growth.

In its paragraph 11, the Addis Ababa Action Agenda commits to "identify actions and address critical gaps relevant" to the 2030 Agenda and the Sustainable Development Goals "with an aim to harness their considerable synergies, so that the implementation of one will contribute to the progress of others." The Agenda has therefore identified a range of cross-cutting areas that build on these synergies.

Among these cross-cutting areas, paragraphs 15 and 16 of the Addis Ababa Action Agenda, respectively, focus on "promoting inclusive and sustainable industrialization" and on "generating full and productive employment and decent work for all and promoting micro, small and medium-size enterprises."

Prior to the 2030 Agenda for Sustainable Development and the Addis Ababa Action Agenda, the relevance of inclusive and sustainable industrial development as the basis for sustainable economic growth was also addressed by the Lima Declaration: Towards Inclusive and Sustainable Industrial Development, adopted in December 2013.

Paragraph 2 of the Lima Declaration reads: "industrialization is a driver of development. Industry increases productivity, job creation and generates income, thereby contributing to poverty eradication and addressing other development goals, as well as providing opportunities for social inclusion, including gender equality, empowering women and girls and creating decent employment for the youth. As industry develops, it drives an increase of value addition and enhances the application of science, technology and innovation, therefore encouraging greater investment in skills and education, and thus providing the resources to meet broader, inclusive and sustainable development objectives."

The mutually reinforcing relationship between social and industrial development and the potential of industrialization to promote, directly and indirectly, a variety of social objectives such as employment creation, poverty eradication, gender equality, labour standards, and greater access to education and health care was also identified in Chapter II of the Johannesburg Plan of Implementation.

Agenda 21 and the Rio Declaration on Environment and Development provide the fundamental framework for policy discussion and action on matters related to industry and sustainable development. Although the role of business and industry, as a major group, is specifically addressed in chapter 30, issues related to industry and economic development, consumption and production patterns, social development and environmental protection cut across the entirety of Agenda 21, including its section 4, Means of implementation.

Note: Explainer 6 is long. Yet I deem it necessary to quote SDG 9 in full to give you a full appreciation of the message behind the text.

2.9 Summary

This chapter defines innovation as a new tangible thing and innovating as a circular repetitive action. The required mindset and skill sets of people working in a company depend on the maturity of the business. Start-ups look for human qualities like flexibility, resilience, perseverance, an open attitude toward opportunities, creativity and a "do-it-yourself" mentality. Established businesses look for control over operation, management of risk, strict adherence to safety regulations and compliance with procedures. It is important to realize that mind comes before matter when you initiate innovation. There is significant power in having the mental skill to imagine other routes to products. Fixed company memories, company memories lost and copied behavior can affect your innovation initiative. You are advised to be aware of these phenomena and to take them on board respectfully. You can use powerful tools to spark dynamic interaction in a group of colleagues who know each other for years: Group Photo, Role Play and Map. If you plan to instigate recurring innovation either you can embark on an already existing Business Evaluation Cycle by extending its scope to Process Evaluation or you can build an argument to initiate it, anticipating increasing outside pressure on continual improvement of products and manufacturing processes, such as Sustainability Benchmarking.

3 The Self-Innovation Method

Abstract: This chapter describes the preparation, explanation and execution of the 6-Steps Self-Innovation Method (6-steps SIM) integrated in a 2-day Workshop format. Preparation comprises 11 different topics, which need to be ready before the Workshop can take place. The two most important ones are the Company Support Base and the Company Innovation Team. An example is given to show how the 6-steps SIM works in practice. The execution part of the chapter describes in great detail in which order, how and with which presenters you execute the 2-day Workshop. On a meta level, comments are continually given throughout the text to help you understand why certain things are done. The chapter wraps up with the final deliverable: a Shortlist of innovative pieces of equipment that enable a step-up in sustainability of your manufacturing process.

3.1 Introduction

In Chapter 1, we have explored your company's environment. We have seen it can be a complex mix of work routines, expected and unexpected, urgent and nonurgent, important and unimportant matters. The direction of the company is affected by internal/external drivers and management trends. Ambiguity with respect to the company's direction can occur when the company Narrative and daily practice diverge. Many of these aspects can hamper your effort to initiate sustainability projects. We have outlined strategies you can use to find out how sustainability can add lasting value to your company. We have listed actions you can take to be effective in such situations.

Chapter 2 shines light on innovation and sustainability. The available mindset and skill sets of people depend on the maturity of the company. Usually start- and scale-ups have more people with innovative mindset and skill sets than companies running mature businesses. We discussed ways to spark and promote an innovative mindset in groups of people who know each other already for a long time and who work with mature manufacturing processes.

This chapter explains the three steps of the Self-Innovation Method in great detail: Preparation, Execution and Follow-up. The ultimate objective of 6-steps SIM is to arrive at a ranked list of new pieces of equipment. This equipment will enable the step-up in sustainability of your process.

https://doi.org/10.1515/9783111383668-003

3.2 Preparation

The following set of items needs to be in place before you can kick off the Self-Innovation Method:

3.2.1 Company Support Base

3.2.2 Presenters

3.2.3 Company Innovation Team

3.2.4 Agenda with timing

3.2.5 Content

3.2.6 Criteria

3.2.7 Brain food and drinks

3.2.8 Venue

3.2.9 Time

3.2.10 Presentation and recording tools

3.2.11 Photo booth/Role Play

3.2.1 Company Support Base

In Sections 1.6–1.8, we have outlined a strategy you can use to build a Company Support Base. As we have stated in "How to position this book" we believe this strategy will enable you to effectively build this Support Base. Rather than reiterating, this paragraph stresses the importance of securing the Support Base before embarking on the 6-steps SIM Workshop. You have to be convinced that your proposal adds a real and lasting value to the company; there has to be buy-in from your peers and higher-ups and you need to have a budget, both for execution of the Workshop and for the next step: Chapter 4. This comprises the development of the Plant Revamp Proposal directly after the Workshop.

3.2.2 Presenters

It matters a great deal who you are inviting to speak during the Self-Innovation sessions. Since your sustainability proposal ultimately will be beneficial for the business, your Business Unit Director should open the first session. He has to lead by clarifying how an increase in sustainability will positively impact the business. His presence during all sessions enhances the sense of importance and motivates Team members.

Equally important on the list of presenters is the internal Champion of the project. Assuming this is you, you are the driving force of this initiative. This is the moment to make clear why you want to pursue this, why you have chosen this Team of colleagues and how their success will contribute to the company. You can also clarify which role you have from start to finish and how people can rely on you during the whole process. You can use this opportunity to show trust, inspire and spark enthusiasm. This is going to be exciting!

The third important presenter is the technologist. You should carefully select, since she/he is going to lead the Team through the layout of the manufacturing process (Table 4). The technologist has to be knowledgeable about the process and skillful in presenting. She/he also has to be able to detect and respond to "weak signals." Weak signals comprise small signs of disagreement, deviating views and different analyses. Having a radar to pick up these signals and discuss them in the group contributes to a commonly shared vision about the manufacturing process. Resolving these signals in an early phase takes time, but it speeds up the next steps.

3.2.3 Company Innovation Team

Upfront selection and discussion with Team members you obviously have to do prior to the Workshop. Giving people an upfront impression of what is going to happen and how they can contribute is motivating and reassuring at the same time. This approach also offers you the possibility to neutralize any arguments not to participate.

We stressed the importance of Team diversity (Sections 2.2 and 2.5). We recommend to invite people like (chief) operators, technologists, (assistant) plant managers, process researchers, maintenance personnel and Lean Six Sigma Black Belt people. The latter category may seem odd. However, next to having useful knowledge about plant performance, Black Belters are very fast in assessing large amounts of data. They can quickly categorize these to create an overview and make things insightful. This is very helpful once the team has generated a long list of alternatives.

Team size depends on company size and the size of your manufacturing process. Start- and scale-ups typically use Teams of four to five people. Multinationals use Teams up to 15 people (Figures 6–8).

3.2.4 Agenda with timing

Trivial as it may seem, having a realistic yet ambitious and detailed agenda including time slots for each individual item is hugely important. The agenda needs to list all steps with an explicit estimate of the amount of time it takes in minutes to complete each step in a satisfactory manner.

⚡ It is essential that the agenda is visible and readable for everybody at all times during all sessions. Putting it on a central spot on a wall in the Session room(s) works quite well.

Table 4: Agenda of 2-day Self-Innovation Workshop.

No.	Timing (h)	Action	Who?	Duration (min)
1	8.00–8.20	**Group Photo** in photo booth	All	20
2	8.20–8.50	Team Member introduction/using three questions	All	30
3	8.50–9.20	Opening by Business Unit Director	Business Unit Director	30
4	9.20–9.50	Opening by Project Champion: Purpose, Rules	You	30
5	9.50–10.35	Technical presentation of Manufacturing Process	Technologist	45
6	10.35–10.50	Break	All	15
7	10.50–11.10	Scope Proposal and Discussion	You/All	20
8	11.10–11.30	Listing Process Bottlenecks	All	20
9	11.30–12.00	Criteria Proposal and selection	You/All	30
10	12.00–13.00	Lunch	All	60
11	13.00–13.50	**Map** group exercise	All	50
12	13.50–14.20	Screening Bottleneck list/selecting one	All	30
13	14.20–14.45	Linking bottleneck to a piece of equipment	All	25
14	14.45–15.00	Identifying process function of equipment	All	15
15	15.00–15.30	Break	All	30
16	15.30–16.00	Identifying the underlying driving force	All	30
17	16.00–16.45	Generating other driving forces that drive the same process function	All	45
18	16.45–17.00	Wrap-up day 1	You	15
19	17.00–22.00	Sports/walks/diner	All	300
1	8.00–8.15	Presenting results of day 1 + Agenda of day 2	You	15
2	8.15–9.15	**Role Play** group exercise	All	60
3	9.15–10.15	Translating process function back into equipment applying the new driving force	All	60
4	10.15–10.50	Celebration; break		45
5	10.50–12.00	Presenting individual results to Team	All	70
6	12.00–13.00	Lunch	All	60
7	13.00–14.10	Presenting individual results to Team	All	70
8	14.10–14.35	Break	All	25
9	14.35–16.41	Ranking equipment using selection criteria	All	126
10	16.41–17.05	Break	All	24
11	17.15–18.00	Classifying equipment in 3 TRL ranges	All	45
12	18.00–18.15	Wrap-up of 2-day session	Business Unit Director/you	15

Assumptions: Team = 7 people; 3 equipment proposals each; 6 criteria; 1 min discussion per proposal/criterion combination.
© H.N.Akse 2025

3.2.5 Content

Different types of content are needed in the sessions. These have to be prepared in advance in chronological order:
- Business presentation/target setting
- Self-Innovation Method presentation + agenda
- Technical presentation of the manufacturing process/scope definition. These include the chosen part of the process flow diagram of the manufacturing process; a heat balance over this part; a mass balance over this part; a brief description of the main processing units in this part; utilities (electricity, steam, cooling water, auxiliary chemicals used in this part).
- Reference syllabi covering intensified technologies and websites of the top 20 process engineering companies serve the purpose of reference works and can be used as sources of inspiration. Chapter 5 covers this subject.
- Criteria document; see Section 3.2.6.
- Original process design books dating back to the time of start-up of the manufacturing process, if available. Preferably these books include design rationales behind the process layout chosen. Often these design rationales are very helpful in understanding why the original process designers made certain choices.

3.2.6 Criteria

At some stage during the Self-Innovation session, the Team will have to rank all innovative options to be able to identify the top 3. To show this is not a trivial thing, assume a Team of 7 people comes up with 21 process alternatives. You can rank these by scoring each process alternative against predefined and preagreed criteria. Assuming a reasonable set of 6 criteria, you already end up with a matrix of 126 fields that have to be scored. Any additional criterion adds another 21 scores to the list. My advice is to compile a draft list of maximum 8–9 criteria upfront, discuss them during the Self-Innovation session with the Team, funnel them to 6 and use them for scoring. You can find a nonexhaustive long list of criteria in Table 6.

3.2.7 Brain food and drinks

This may seem a peculiar item on this preparation list. However, it is a serious thing and closely linked to Section 3.2.8. Experience shows that during the Self-Innovation sessions, there are certain time slots of 30–45 min during which everyone's concentration level is very high and brains run at maximum capacity. A room temperature increase of up to 4 °C is not uncommon! This physically requires lots of fuel. The best brain foods contain large amounts of fructose and glucose in a form that is easily absorbed by the human body. To compensate for these "unhealthy" sugars, you may want to add lots of fresh fruits like apples, bananas, grapes and various sorts of nuts. Fruit shakes generally go down very well, as well as soft drinks, herbal teas, etc.

3.2.8 Venue

The venue preferably is *not* on the company's premises. What works is a remote location near a sea, a lake or a forest (Figure 11). This makes people feel special and reduces the chances of them being called away for urgent business or operational issues at the plant.

Figure 11: Royal Palace Het Loo, Apeldoorn, the Netherlands. The 6-steps SIM's first deployment was only a golf club shot away.
Image 2025 ©H.N. Akse

The rooms you use are equally important. Section 3.2.7 explained the impact of putting many brains to work. It means oxygen supply is key! This requires good ventilation and high ceilings. The availability of proper chairs and comfortable sofas in the main meeting room or in a room next door pays off. You cannot ask people to sit on unclad wooden furniture for hours on end.

3.2.9 Time

Time has two sides to it. On the one hand, you need to give the people time to start up. Self-Innovation is new for everybody and people need time to adapt to their roles. Also, the creative part of the process needs sufficient time. You cannot say "be creative" in the same manner as you cannot say "be spontaneous." On the other hand, pressure enhances the quantity and quality of output. Given too much time, people get bored and start doing other things. An additional argument for setting a time limit is the fact that your colleagues are taken out of their daily jobs. Others have taken over their work temporarily. Hiring venues and conference rooms, arranging over-night stays and consuming food and beverages generate real out-of-pocket costs.

From experience I have learned, a 2-day period is the best compromise consider-ing all factors involved. One of these days can be a Saturday, which reduces cash-out cost and gives people the chance to show their commitment at the same time.

3.2.10 Presentation and recording tools

This is technical stuff and most of the time we consider it to be unimportant. Our culture has adopted laptops, beamers and interconnects up to a level where we say: it should just work. Have I got news for you! Most of the time IT still does not work at the moment you really need it. It did not work reliably in 1994 when Personal Computing conquered the world. It still does not work reliably today, more than three decades down the road. Just check your memory to find out how many times you have been part of an audience waiting for the presenter to get a proper image on screen. Think about the cumulative amount of time wasted! These IT manufacturers only have one job . . .

So here is my message: make sure all presenting and recording equipment works; don't assume. Check and recheck. And then check again. Create **redundancy** for all critical components.

3.2.11 Wrap-up

You need to think in advance how you are going to process and present all outcomes *during* the Self-Innovation sessions. Having a continuous overview is essential for ev-eryone. My advice: use a laptop/PC combined with a high-resolution beamer. Advan-tages of a laptop/PC over flip-over sheets or wallpaper are better readability and the opportunity to quantify outcomes by adding score numbers. For example, if you score a process alternative against a criterion like "CO_2 emitted," it can range from 5 (= very good; no CO_2 emission) to 1 (= very poor; no reduction in CO_2 emission). Scores and criteria can be changed and rediscussed while keeping the readable overview pro-jected with the beamer until everybody is happy with the end result.

© Olena Bohovyk - Unsplash

3.2.12 Photo booth/Role Play

Details have already been discussed in Section 2.4. The booth should be set up in a separate room. Remember to have this installed on the evening before the start of the first Self-Innovation Workshop. Again, check and recheck your equipment. Make sure the photo booth room is large enough to also accommodate Role Play.

3.3 Execution

First, we show you a brief version of 6-steps SIM using Figure 12. This is followed by a constructed example of how SIM works in practice. Then we present a detailed chronological description of every step in the Innovation Workshop Session, including all items described in Section 3.2. We wrap up by summarizing all chronological steps in a detailed workflow diagram shown in Figure 19.

3.3.1 6-Steps SIM at a glance

SIM starts with a list of process bottlenecks identified by the Team within the agreed Scope Figure 12. Most of the time, a bottleneck can be linked to a piece of equipment (1). What is the process function of this equipment (2) (see Table 7)? What is the underlying driving force that makes this process function happen (3) (see Table 8)? Are there other driving forces that can drive the same process function (4)? Can the process function be translated back into a piece of equipment using this newly found driving force (5)? Can you classify this new piece of equipment in TRL terms (6)?

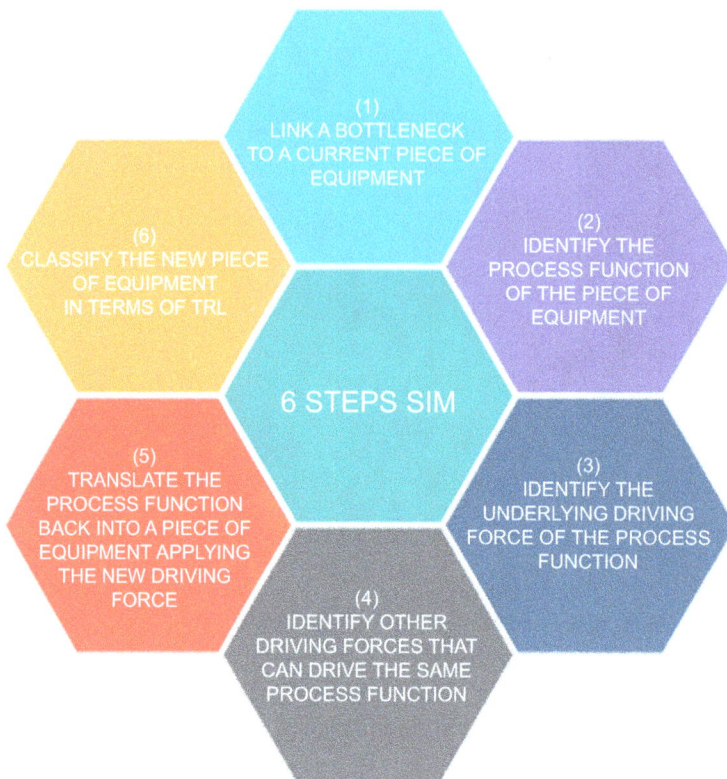

Looking at this sequence of steps you will notice elements that have been discussed before:

- You start on equipment level (1) and you end on equipment level (6).
  This is in line with our definition of innovation in Section 2.1: innovation is a new thing
- The in-between steps gradually get abstract (3, 4) to finally arrive on a level where you think laterally/bisociatively in terms of fundamental physical and chemical forces, leaving the level of thinking in equipment terms. On this most abstract level, the real change is found. Subsequently, you gradually get concrete (4, 5) to arrive at the equipment level again. This is in line with Section 2.3: "the mental exercise of creation and innovation always precedes design and construction of a plant."
- You are not "short-circuiting" by simply exchanging one piece of equipment by another. This is common practice in industry when there is an issue with a piece of equipment. In itself, this is perfectly understandable. You look for a new device with the same specifications by calling a few vendors on your preferred vendor's

list. From the quotations you get, select the one with the best price/performance ratio and that's it. Most of the time, the new equipment is a few percent better in efficiency/capacity/energy, which is a bonus. Nice, but this is not going to give you the required step up in process sustainability.

It is important you see and fully understand the move we are making here, from concrete to abstract and then back to concrete again. Once you have a mental picture and a real feel of this move, it will be easy for you to understand the type and sequence of actions described in Sections 3.3.4. Section 3.3.2 starts helping you in creating this mental picture.

3.3.2 SIM example: security of drinking water supply

This paragraph purposes to give you a concrete impression of how SIM works. It is a constructed example since all real SIM outcomes are under NDA; therefore, these cannot be made available in the public domain.

Let us assume you bought a house some 15 years ago in the beautiful town of Laukanlampi (Finland), a 40-min drive from Joensuu. Close to Laukanlampi, the Uimasalmen bridge separates the Rukavesi lake from the Rahkeenvesi lake. From Finland's capital Helsinki to Joensuu, it is a 5-h drive if you take the E75 via Lahti and Heinola. Laukanlampi is situated in a beautiful water-rich area. Although all houses were heated with waste heat from the nearby paper mill (Figure 14), the small town did not have a central drinking water supply. The first inhabitants had solved this by installing a standard water distillation skid that simply took in seawater and removed salt and other minerals, thus producing drinking water (Figure 13). It only required a heat source slightly over 100 °C, which was the temperature level of the same waste heat source used for domestic heating.

Over the years, Laukanlampi became really popular, attracting more and more people who came to live and work there. The municipality issued building plots as there was ample space available. The paper mill saw an increasing number of customers for their waste heat. This went on in a satisfactory manner until this development at some point came to a grinding halt. All of the paper mill's waste heat was now in use!

This presented the municipality with a problem as the number of new inhabitants continued to increase. A registered SIM consultant was called. He came all the way from Helsinki. Right at the onset of his first meeting with representatives of the municipality this gentleman started shooting questions at two benevolent civil servants – like any decent consultant would. You will recognize the 6-steps SIM shown in Figure 12 in the following Q&A described in Table 5. Please notice, the civil servants of Laukanlampi appear to be particularly well-versed in the subject. They even use proper chemical engineering jargon.

Figure 13: From a small-scale waste heat-driven drinking water unit . . . .
© 2011 Ebbe Holsting

Figure 14: Part of the Laukanlampi paper mill.
© 2024 Aron Yigin / Unsplash

The remaining questions for the municipality concerned plant sizing, distribution and associated investment. Did they go for one central RO unit supplying water to all

Table 5: Example of 6-steps SIM Q&A.

Question of registered SIM consultant	Answer from Laukanlampi civil servants
What is this process about?	It makes drinking water out of salt water using waste heat from the paper mill near our town
Is there a bottleneck?	Yes, the paper mill has contracted all of its waste heat, so there is nothing left for our new inhabitants
1 Can this bottleneck be linked to a piece of equipment?	Yes, the equipment used in all houses so far is a compact water distilling unit running on waste heat from the paper mill
2 What exactly is the process function of this equipment?	Separation of minerals from water
3 What is the current driving force?	The current driving force is the difference in volatility between water and minerals at water boiling temperature at 1 bara
4 Can you identify other driving forces that can drive the same process function?	The process function is separating minerals from water. Other driving forces that can do this are: – Electrical voltage gradient between two electrodes in combination with ion-specific membranes – Relative permeability in a polymer substance in combination with a mechanical pressure difference – Relative adsorption speed on solid surfaces – Etc. (nonexhaustive list)
5 Can you translate the process function back into a piece of equipment applying the new driving force?	Reverse osmosis membrane process uses the permeability difference of water molecules and mineral ions in a polymer substance in combination with a mechanical pressure difference generated by an electrically driven pump
6 What is the TRL level of the new piece of equipment?	Highest: RO units come in all commercial sizes

© 2025 H.N.Akse

houses, both existing and new? (Figure 15) This would involve construction of an all-new town-wide drinking water infrastructure as well as a dual pipeline system from the sea to the RO plant to supply seawater and remove brine. Or would the municipality set up a funding program enabling inhabitants to install small-sized domestic RO units in their houses? This would require only new inhabitants to arrange for dual pipeline systems from their houses to the seashore. Whatever the town decided in consultation with the existing and new residents, the issue of security of the supply of drinking water was solved by applying the 6-steps SIM.

3.3.3 Workflow: day 1 of SIM Workshop

Traditionally, SIM kicks off with Group Photo to open up the minds and to spark dynamic interaction in the Team. Directly afterward, every Team member introduces himself/herself by answering three questions: job type, number of years with the company and what makes your job tick – what ignites your enthusiasm? From this sequence, it is already clear we put the Team first, the rest next. You can use the input from the Team to already make links to the next steps of the session. This triggers curiosity among the audience. You can also do the "total year count" by adding up the number of years every Team member has worked for the company. This quantifies the cumulative experience and process know-how of your Team (Section 2.2). Most of the time you'll find the sum of years is surprisingly large.

Since ultimately the sustainability enhancement of the process will be beneficial to the business, the Business Unit director ultimately responsible for the manufacturing process will open the session. He will present the expected impact of sustainability on future volume of sales and an increase in the number of markets. He wraps up with sustainability targets arising from the business.

Next, the SIM Champion explains SIM in a nutshell. Assuming this is you, you also show the agenda of day 1 comprising actions + timing. You also explicitly repeat the rules of trust at this stage. What happens in the room stays in the room. During brainstorming phases, "stupid proposals" simply do not exist. Critical scrutiny is postponed to the selection phase.

What follows is a technical presentation of the manufacturing process by a knowledgeable technologist. With the use of a process flow diagram, the process is explained from feedstock to product. The history of the process is discussed. Start-up date, initial name plate capacity, extensions, work-arounds, major product specification changes, current annual capacity, etc. are presented. This gives the Team insight into changes that have been made to the original design. The presentation includes a listing of relevant quantitative data like mass flows, specific utility use (power, steam, cooling water), emissions and use of utility chemicals. The technologist wraps up with a translation of the business targets into SMART[4] technical targets: steps down in energy use, emission of greenhouse gases, cooling water use and specific feedstock consumption.

Third, you present a Scope proposal. This entails limiting the focus to a certain part of the manufacturing process. The Team gets to challenge this Scope, discuss it and make changes until there is consensus. At this point, everybody is convinced the chance of a step-up in sustainability is highest within this part of the plant.

Fourth, the Team discusses whether there are any process bottlenecks within the Scope. We do this since often these bottlenecks are leads for SIM. Examples of bottlenecks are: limited conversion due to equilibrium reactions, large recycle, significant amounts of by-product formation, excessive fouling, high specific energy use of a unit operation, high viscosity of a process stream, undesired temperature excursions in a reactor, buildup of trace components in a distillation column, limitations in heat and/or mass transfer, large pressure drops, difficult filtrations and long settling times of fines.

Companies always find work-arounds that successfully deal with bottlenecks. Although these solve the bottleneck, work-arounds always come at a cost. This may take the form of frequent replacement of filters, a high frequency of fouling removal, intensified maintenance of wearing parts, daily cleanout of heat exchangers, the existence of large recycles of liquids and/or gases. These phenomena in turn often translate themselves into higher energy and/or chemical consumption to overcome larger pressure drops, additional downtime of parts of the plant, extra operator attention for blocking-in/blocking-out procedures, cleaning procedures, etc. In all instances, the additional cost incurred can be carried by the margin made, so process economics is OK. By definition, otherwise it would not really be an effective work-around, would it?

The fifth item on the list is the set of criteria the Team is going to use to rank all process alternatives that will be generated. To kick-start this discussion, you present a set of proposed criteria. Table 6 lists some general criteria you can use. My advice is to define and select criteria tailor-made for your process.

The purpose of criteria is to distinguish between process alternatives found. Scoring each process alternative against a set of criteria enables ranking. Ranking helps in discovering the best alternatives. Finding and agreeing on good criteria takes time. Oftentimes, there appears to be one criterion yielding the same score for each alterna-

---

4 SMART: Specific, Measurable, Achievable, Realistic, Timely.

Table 6: General ranking criteria (nonexhaustive).

1	Boosting process efficiency
2	Combining unit operations
3	Enhancing conversion per pass
4	Generating a sharper particle size distribution
5	Improving yield in reactor section
6	Lowering operational cost
7	Lowering specific fixed capital
8	Minimizing plant footprint
9	Minimizing plant size
10	Minimizing plant weight
11	Minimizing recycles in number and size
12	Minimizing specific steam/cooling water/electricity consumption
13	Reducing CO_2 emission
14	Reducing impurity levels in the final product
15	Reducing number of unit operations
16	Splitting one unit operation into two with different functions

tive when used in a ranking session. Obviously this one should be left out and replaced by another one. This is why you should aim for a set of eight to ten criteria at the end of the Team discussion. Then you have a few spare ones if one or two criteria appear to be non-discriminating during the ranking exercise. A final set of six criteria during the ranking session is optimal (§ 3.2.6).

At this point, the Team gets some time to relax. We then continue with the sixth item: a team exercise called Map (explained in Section 2.5). This further loosens up the traditional relations inside the Team and sparks enthusiasm.

After this plenary part, we start with the seventh part. This is the first of two parts that will be carried out individually by each Team member (about individuality: Explainer 7). This is going to be a high-intensity piece of work. It needs oxygen, edible fuel (Sections 3.2.7 and 3.2.8), a spacious room, flip-over sheets and markers and a time maximum of 30–45 min. The flip-overs can be used to write down ideas for later presentation to the Team by each Team member.

Each Team member first tries to answer the question: the removal of which bottleneck will most likely meet the sustainability goals set? Once found, the Team member then identifies a link between the bottleneck and a piece of equipment in the Scope (Figure 12, step 1).

The next question then is (Figure 12, step 2): what exactly is the process function of this piece of equipment? Table 7 gives a nonexhaustive list of process functions.

i **Explainer 7: individual or collective brainstorm?**
During SIM Workshops, people many times argue parts 1–5 in Figure 12 should be done collectively. Most of the time their argument is that in their company they are used to do creative work like brainstorms in a group. Sometimes they are even reluctant to kick off individually.

There are two reasons why you should stick to the individual approach. The first one is appreciation of the unique combination of experience, mindset and skill sets of each Team member in your group. Unique combinations may well lead to unique proposals! So you better give it a chance. Reluctance can be a way to hide insecurity or lack of self-confidence. If this is the case, these things need to be dealt with in a way other than changing the brainstorm to a collective exercise.

The second reason to stick to the individual approach is the very fact that people are used to group-wide brainstorms. In fact, they state this explicitly. That way, they remain in their comfort zone. Responsibility for results gets fuzzy when you are part of a group effort. Many times, a well-articulated usual suspect starts dominating the content and direction of the brainstorm, leaving behind a timid group of relieved and frustrated followers: relieved since they did not have to take responsibility and frustrated since they did not speak out their own mind. This reduces a true brainstorm to a parade of one leader and many followers. We won't have that.

Table 7: Examples of process functions (nonexhaustive).

1	Absorbing	22	Decharging	43	Heating		
2	Adsorbing	23	Diffusing	44	Impulse transferring		
3	Blowing	24	Disintegrating	45	Isolating		
4	Breaking	25	Disperging	46	Liquefying		
5	Catalyzing	26	Dissolving	47	Mass transferring		
6	Centrifuging	27	Distilling	48	Melting		
7	Charging	28	Distributing	49	Milling		
8	Coagulating	29	Dividing	50	Mixing		
9	Collapsing	30	Dripping	51	Permeating		
10	Collecting	31	Drying	52	Pouring		
11	Combusting	32	Emulsifying	53	Pumping		
12	Compressing	33	Evaporating	54	Reacting		
13	Condensing	34	Expanding	55	Separating		
14	Conducting	35	Extracting	56	Shearing		
15	Convecting	36	Extruding	57	Solidifying		
16	Converting	37	Filtrating	58	Spinning		
17	Cooking	38	Flashing	59	Spraying		
18	Cooling	39	Foaming	60	Storing		
19	Crushing	40	Freezing	61	Stripping		
20	Crystallizing	41	Gelling	62	Sublimating		
21	Cutting	42	Heat transferring	63	Wetting		

Note: This set of process functions recurs in Appendix 2 in an integrated format together with driving forces and maturity levels for each of 81 different intensified chemical technologies.

Having identified what exactly the process function is, the next question is (Figure 12, step 3): what is the driving force that makes this process function work? Table 8 gives a nonexhaustive list.

Table 8: Examples of driving forces (nonexhaustive).

Energy field		Force field	
1	Electric	10	Affinity
2	Electromagnetic	11	Centrifugal
3	Electrostatic	12	Concentration
4	Free Gibbs	13	Gravity
5	Laser	14	Impulse
6	Light visible	15	Magnetic
7	Microwave	16	Permeability
8	Plasma	17	Pressure
9	Temperature	18	Shear
10	Ultrasound	19	Solubility
		20	Velocity
		21	Volatility

Note: This set of driving forces recurs in Appendix 2 in an integrated format together with process functions and maturity levels for each of 81 different intensified chemical technologies.

At this point, you have identified a limiting piece of equipment. You know its process function and you know the driving force that makes it work. The next question every Team member asks himself/herself is:

> **?** **Do I know other driving forces that drive the same process function?**

From experience I have learned that this is the most powerful question you can ask during the 2-day session. It triggers a lot of creativity since it opens up mental windows to new ideas. Nobody speaks. Everybody writes large amounts of text on their flip-over sheets to stick these to the wall. Room temperature goes up. Without exaggeration I can state this is the pinnacle of your innovation session. For you this means you have to do everything to make this a success. Your effort may range from answering questions to controlling fresh air and room temperature to encouraging people who find it difficult to go with the flow. You push every button to make it work.

At this stage, we are at the end of part 7 of the session that has been carried out by everybody individually. The Workshop will also have reached the end of the first day. This is a natural moment to wrap up.

You do this by summarizing the results of each of the seven parts of day 1. You also give a brief outlook on day 2. This ends the official part of the day.

I highly recommend you to spend the evening together as a group. You may consider sports, a nourishing meal, a healthy outdoor walk in the woods/along the beach (Figure 16). Try to do stuff unrelated to the work you have been doing all day. This offers your Team members the chance of recuperating both mentally and physically. The other thing that will happen is of a more psychological nature. Everybody's brains will keep on processing the events of the day. On a more subconscious level, ideas will be formed and shaped. This takes time and a good night's sleep. All of this helps to be successful on day 2 of the SIM Workshop.

Figure 16: A healthy outdoor walk.
© H.N.Akse 2025

3.3.4 Workflow: day 2 of SIM Workshop

On day 2, you kick off reiterating what you told your Team at the end of day 1. This brings back to mind all learning points. Next, you present the agenda items including timeslots for the day that lies ahead. Before embarking on part 8 of the Workshop, the Team executes "Role Play." Purpose and impact of this exercise – sparking enthusiasm and creating dynamic group interaction – have been explained in Section 2.5.

After Role Play you carry out the second and last individual part of the Workshop. Having identified the most suitable driving forces that can drive the same process function, the next question is: can you translate this function back into a piece of equipment applying the new driving force? This question also generates lots of creativity. For you the same rules of engagement apply as in the afternoon of day 1. Again, you have to do everything to make this a success. Your effort may range from answering questions to controlling fresh air and room temperature to encouraging people who need reassurance and some initial guidance to find their way.

Each Team member draws his/her newly identified piece of equipment on a flip-over sheet. Items that need to be on the flip-over sheets are given in Table 9.

Figure 17: Show your ideas!
© Yolk - coworking Krakow - Unsplash

Table 9: Items needed on flip-over sheet for each novel alternative.

- Currently limiting piece of equipment
- Limiting process function
- Current driving force
- Driving force alternatives that can drive the same process function
- Newly found piece(s) of equipment that can deploy the new driving force, enabling the same process function
- Indication of process conditions (temperature, pressure, concentration, residence time, etc.) of the new piece of equipment
- In- and outgoing mass and heat flows
- Sketch of internals
- Indication of TRL level

Taking into account the normal human attention span, this part of the Workshop can last for 30–45 min max. When ready, all flip-over sheets are put on display on the wall with adhesive tape.

Next item on the list is a long break of about 45 min. It is used for recuperation, brain food and drinks, taking in fresh air, stretching legs. Each Team member prepares a brief presentation for the Team about his/her findings.

From here on, it is a group exercise again. Each Team member presents his/her findings to the Team using the flip over within a time span of 10 min max. The purpose is everybody gets a proper understanding of the underlying reasoning of the innovative new piece(s) of equipment on display.

My advice is to encourage people to ask clarifying questions. Depending on group size, this may take between 1 and 2 h. When time exceeds 1 h, I advise to take a 10-min break. From an operational point of view, it is important someone keeps track of each proposed alternative in an Excel sheet. Table 10 gives an example. The minimum set of data that needs to be recorded per alternative comprises the items printed in blue in Table 9.

At this point, you and your Team have produced a long list of innovative pieces of equipment. It is worthwhile mentioning this explicitly to your Team. A major piece of work has been done and everybody deserves kudos for this achievement! Actually, this is a nice opportunity to celebrate this on the fly, e.g., by ordering special teas or coffees with pastries or fruits, whatever the Team appreciates. You can make this special by having hotel staff serving the delicacies inviting the cook to come over to the Workshop and explain how everything is freshly made.

From a helicopter point of view, until now we have been diverging by generating more and more alternatives. From now onward, we start converging toward a shortlist of alternatives that have the highest chances of success. This is where the criteria come in that have been defined by the Team (Sections 3.2.6 and 3.3.2). Assuming six criteria, a Team of seven people and three alternatives per person, you are looking at a matrix of $6 \times 21 = 126$ options in Excel that need scoring.

There are two ways of processing this matrix. My experience covers both ways. One way is to have the Team thoroughly discuss each option and put down the resulting score as a number in Excel. If an alternative matches well with a criterion, you give it five points. If an equipment alternative does not meet a criterion at all, you give it one point. If the Team feels that one or two criteria are more important than the others, then you can easily give these a weighing factor >1, leaving the others at 1.

Below the yellow headers, you put the proposals as shown in Table 7 on the left. On the right (green) you notice a set of six criteria agreed upon by the Team. Each criterion has its own weighing factor. Row by row you briefly discuss how a proposal scores on each criterion until you have covered all 126 items.

However tough, time-consuming and old school this may look, it can be done by spending 1 min maximum per option. The "Excel route" in this example requires 146 min for processing of 126 items. This approach poses a huge challenge to the Team. It demands strict discipline and sustained concentration, but you get value for money. No item is overlooked. All items are scored. Excel offers the opportunity of changing scores or weighing factors afterward if the Team deems this necessary. And importantly, since everybody can deliver input and has an overview at all times, you create unanimous support for the end result.

Table 10: Example of 6-steps SIM results taken from a real Workshop.

No.	Bottleneck	Equipment	Function	Current driving force	New driving force	Function	Equipment	Benefit
1	Only half of the substituting agent is used	Reactor	React feedstock and oxydator	Substitution of feedstock	Start with monomer and polymerize	Polymerization	To be identified	Eliminate the by-product
2	Volatility of oxygen	Reactor/vacuum system	Oxygen removal from slurry	Relative volatility	Mechanical force	Oxygen removal from slurry	Reactor/ultrasonic device	Faster and more effective air removal
3	Removal of by-product	Washing system	Washing of feedstock	Use of gravity and belt filter	Gravity, pressure	Extra floor, belt filter washing, belt filter press	Belt filter, belt filter press	No pumps, no neutralizer
4	Feedstock transfer speed	Feedstock transfer line	Feedstock transfer to slurry make-up tank	Pressure difference	Gravity	Feedstock transfer to slurry makeup tank	Extra floor, use gravity to transfer feedstock	Decreased maintenance, faster transfer
5	Speed and effectiveness of moisture removal	Dryer	Removal of moisture from feedstock	Volatility of H_2O and by-product	Infrared, ion exchange	Removal of moisture from feedstock	Infrared device added to dryer	Increased speed and effectiveness of moisture removal
6	Removal of moisture	Dryer	Removal of moisture	Volatility of H_2O and by-product; heat transfer	Microwave	Microwave drying	Microwaves	Electricity cheaper
7	Creation of by-products	Reactor	Reaction of feedstock and oxydator	Mass transfer and reaction of the by-product (downstream)	Chemical reaction (in reactor)	Add another chemical into the reactor to eliminate the by-product	No change	Eliminate the need to remove the by-product later in the process

8	Removal of liquid	Acid centrifuge	Removal of moisture	Centrifugal force	Pressure difference	Removal of moisture and washing	Belt or drum filter	Do not need to use the neutralizer
9	Removal of acid	Washing tanks	Removal of by-product	Relative volatility and reaction of by-product	Relative volatility	Repeat washing steps (centrifuge, reslurry, strip)	Similar equipment, just more	No use of neutralizer
10	Removal of moisture	Centrifuge/dryer	Removal of moisture	Centrifugal force, volatility of moisture	Centrifugal force, pressure difference	Take to compounding without completely drying	Centrifuge, mechanical press and pelletizing	Eliminate the need for dryer
11	Neutralization of waste stream	Wastewater system	Neutralization of waste stream	Reaction of by-product with utility chemicals	Blue power (reverse electrodialysis)	Electrochemical by-product neutralization; energy from waste	Electrochemical membrane unit	Electrical energy from waste
12	Removal of moisture	Dryer	Removal of moisture	Steam drying	Freeze drying; freeze water and remove acid	Freeze water after acid centrifuge	Freezer	Eliminate steam usage
13	Mass transfer	Reactor	Reaction of feedstock and oxydator	Concentration difference in liquid	Concentration difference in gas phase	Gas-phase substitution	Gas-phase reactor	Easier removal of by-product, no drying of feedstock
14	Chemical reaction equilibrium; reaction rate	Reactor	Reaction of feedstock and oxydator	Steady-state chemical equilibrium	Unsteady-state equilibrium (pressure leading to concentration differences)	Modulate pressure (increase/decrease)	Reactor	Help to drive the reaction
15	Reaction equilibrium	Reactor	Reaction of feedstock and oxydator	Mass transfer	Improved mass transfer (concentration differences)	Remove by-product via membrane on recycling of the reactor	Membrane, recycle loop	Improve the reaction rate

(continued)

Table 10 (continued)

No.	Bottleneck	Equipment	Function	Current driving force	New driving force	Function	Equipment	Benefit
16	Removal of acid	Washing tanks	Removal of by-product	Heat, thermal, volatility	Ultrasonic	Force by-product out of pores without heat	Ultrasonic pump with chamber	No exposure of heat to feedstock in the washing process
17	Removal of oxygen from feedstock	Slurry transfer line to reactor	Transfer slurry to the reactor	Pressure/vacuum	Ultrasonic	Ultrasonic removal of oxygen from feedstock slurry	Ultrasonic pump with chamber	More effective removal of oxygen from slurry
18	Only half of the substituting agent is used	Reactor	Reaction of feedstock and oxydator	Photochemical process	Electrophotochemical process	Convert by-product back to oxydator	Electrolysis equipment (in our process)	Increase raw material efficiency
19	Removal of acid	Washing tanks	Removal of acid from feedstock	Heat, thermal, volatility, neutralization	Heat, thermal, ultrasonic	Removal of acid from feedstock	Ultrasonic pump in addition to existing equipment	More effective removal of acid
20	Removal of water	Reactor	Reaction of feedstock and oxydator	Mass transfer through water	Mass transfer through solid CO_2 or liquid CO_2	Use supercritical or liquid CO_2 to help plasticize the feedstock	Supercritical reactor	No water needed
21	Removal of acid	Washing tanks	Removal of acid	Heat, thermal, volatility, neutralization	Relative volatility	Extraction of acid with supercritical CO_2	Extractor	More effective removal of acid; no steam usage in strippers

Note: Specifics have been converted into generic terms for confidentiality reasons.

Table 11: Example of the "Excel" ranking route.

Ranking	Proposal number	Process Function	New Piece of Equipment	Benefit	SELECTION CRITERIA						
					Maintenance cost down	Capital cost down	Conversion cost down	Zero waste	Stability	Flexible production	Total score
				Weighing factor (1 = minimum; 10 = maximum)	1	2	2	1	2,5	1,5	10

© 2025 H.N.Akse

The other way of processing the matrix is to fly in a Lean Six Sigma Black Belt and let him have his way with this pile of alternatives. Black Belts have a knack for quickly recognizing patterns, similarities and odd ones out. They work with Sticky Notes: well-known colored pieces of paper on which each Team member has written the yellow-colored items of Table 9. One Sticky Note for each alternative. Your Black Belt is helped by the empirical fact Team members often come up with identical solutions. What happens in practice is that a preselection is ready before you know it, and the whole body of ideas is structured into six up to ten clusters of distinctive new types of driving forces/equipment. The photos in Figure 18 show the results generated during a real SIM session. In this specific case, the Black Belt processing time was about 30 min.

Figure 18: Example of the "Black Belt" ranking route.
Photos 2025 ©H.N. Akse

Once the clustering is done, the Team gets to discuss which alternative is the best one in each cluster. Scoring by giving points is done in the same way as when using Excel. Points are written on the Post Its. Total scoring time is about 90 min. The main reason this approach takes less time is due to the fact that a lot of undoubling and combining have been done by your Black Belt colleague.

The "Black Belt" route needs about 120 min for processing of 126 items. This route is slightly faster, relies heavily on the personal skills of your Black Belt, offers less overview for the Team at any point in time and is less flexible in making score changes afterward. Of course, this is your call. The important thing is you have to make up your mind well before you enter this phase of the Workshop.

At this stage you and your Team have a Ranked Shortlist of equipment alternatives in your pursuit of enhancing process sustainability.

The final part of the Workshop is reserved for assessing TRL levels of the proposals. Here your objective is important. If you pursue the shortest route to a sustainability project you only need the high TRL alternatives. If you also want new leads for your R&D department you will need a few of the lowest TRL alternatives as well. If you are considering future cooperation with companies or institutes that are already developing alternatives, you will want a few mid-range TRL alternatives. Whatever you choose, it is your call. My advice is to think this through prior to the SIM Workshop.

If you go for the full Monty, there will be three types of proposals:

– The first group has a TRL of 9 and can be bought off the shelf from equipment vendors.
– The second group has a TRL between 5 and 9 and is known to be in a research and development phase at an institute or inside another company.
– The third group has a TRL of 1. This group contains promising concepts of novel equipment that does not exist yet.

We have come to the end of the 2-day Workshop. My advice is to keep your wrap-up brief. Put the Group Photo on a large screen. Stress the fact that the group effort has been indispensable in achieving the results obtained. Highlight the funny and exhilarating moments. Praise all participants for stepping out of their comfort zones. Promise, you will keep everybody up to date with respect to follow-up: design to deploy. Send every participant a nicely framed Group Photo a week after the Workshop, including the names of all participants on the back.

ℹ Explainer 8: How the 6-steps SIM meets company needs and fosters creativity?
This explainer illustrates how 6-steps SIM meets company needs as well as the need for a free mental space that encourages creativity.

On the one hand, you will have noticed that the 6-steps SIM is a well-structured approach. It is the existence of structure in particular that companies appreciate. Generally, the response of companies embarking on an innovation initiative is one of anxiety. Oh man, our people start thinking out of the box. For two whole days. Away from home. Where does this end?

It is difficult for any manager – if not impossible – to approve a project without a properly described route to an unknown end result. How can he or she be sure this is not going to be a haphazard exercise? Two days of fun, fuzzy results and a bunch of dizzy people next Monday. How can he explain this to his boss? This is why the 6-steps SIM delivers the structure. It reassures sponsors that there is a series of actions focusing on concrete deliverables at the end: new pieces of equipment that can be integrated in an existing plant.

On the other hand, SIM creates a free creative space for all Workshop participants, encouraging them to open up, think laterally and associate freely. It also creates an open and safe atmosphere in which people dare to disclose new ideas. This is the nonstructured part of the method where everything is allowed that enhances process sustainability. Still, even in this nonstructured phase, all Team members base their creative process on fundamental physical and chemical principles.

The resulting Workshop structure offers both the company and the free spirits what they need.

Figure 12 shows an overall picture of 6-steps SIM. Figure 19 shows the 6-steps SIM integrated in the detailed workflow of a 2-day Workshop described in this paragraph. You will notice that many more items are involved in the Workshop to make the whole process run smoothly.

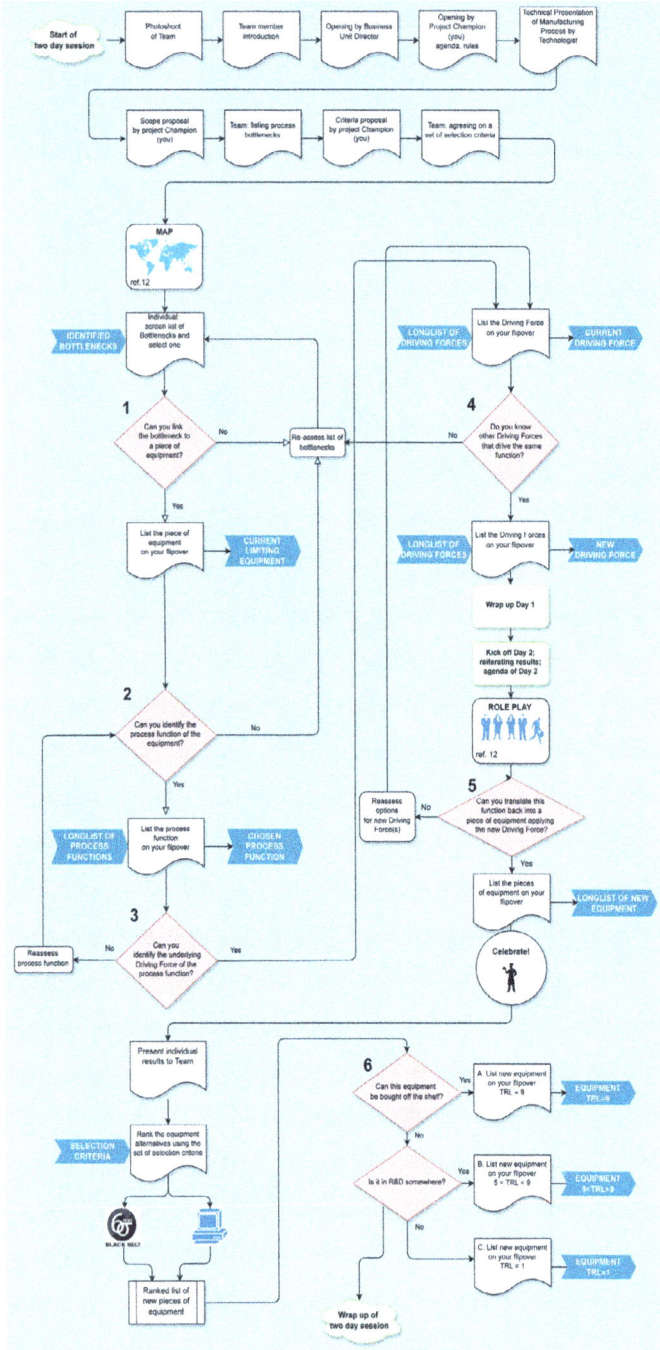

Figure 19: 6-Steps SIM integrated in a 2-day Workshop.
© 6-Steps SIM in detail 2025 H.N. Akse

4 Design to deploy

Abstract: This chapter stresses the importance of planning and executing a design-to-deploy phase directly after the 6-steps SIM sessions. This yields a Plant Revamp Proposal. Assuming that your company has proprietary design methods to develop this proposal in-house, this chapter presents a set of Sustainable Design Guidelines meant to help you in integrating the innovation in the existing plant in a sustainable way.

4.1 Introduction

The Workshop has yielded a series of three to nine process alternatives with TRL's varying from 1 to 9. The core of each alternative is a novel piece of equipment which is not part of the present process layout of your manufacturing plant. Your Team is convinced that this equipment is going to boost the sustainability of the process. Limiting yourself to the high TRL cases, you know that these pieces of equipment can be bought off the shelf from vendors. This is reassuring but it is not the end of the story. Here it makes sense to check in which phase of the sustainability project you are. Referring to Section 1.8 your project consists of four phases:
a)   A Self-Innovation Workshop delivering the best step up in process sustainability
b)   Development of a Plant Revamp Proposal (PRP)
c)   Ranking of all Investment Proposals
d)   Deployment when the Proposal is approved

My advice is to develop the PRP directly after the Workshop. There are two reasons for this. First, your Team is delighted that a solution has been found. Your Team members share the drive to take it further. For you it makes perfect sense to build on this drive. Second, by executing the Workshop you have raised expectations among bystanders: the people you consulted at the beginning of your endeavor; other colleagues who appeared to be supporting your initiative; your manager who has funded the project. They all hear about the positive outcomes. Off course they are curious about the chances of turning this into a successful deployment. Keeping momentum is the adagio.

What can happen when you do not come up quickly with a quantified revamp proposal is curiousness and positive anticipation turn into skepticism. Once this happens, you quickly lose your Support Base. This impedes development of a convincing revamp proposal. Especially when you are in situation 2 of Figure 1 (sustainability is part of the Narrative but not visible in deployed projects) or in situation 4 (sustainability is neither part of the Narrative nor visible in deployed projects). In these two situations your project already was far from business as usual right from the start.

https://doi.org/10.1515/9783111383668-004

When a quick revamp proposal is not forthcoming, bystanders can easily turn into convinced nonsupporters. Before long, their perception of the Workshop has changed from positive and promising into a company-paid luxury event. My guess is, you don't want that to happen since it takes a lot of effort to change that image back into something positive.

The reason for producing a PRP is that you need to have your proposal ranked together with all other Investment Proposals developed in your company. I assume that your company has such a ranking procedure in place like most companies. Table 12 shows what a PRP consists of.

Table 12: Elements of the Plant Revamp Proposal.

- A revised Process Flow Diagram
- Adapted mass and heat balances
- Investment cost estimate (±40%) consists of:
    - Equipment sizing
    - Purchase cost of equipment (using cost engineering algorithms and equipment sizing or dedicated vendor quotes)
    - Total physical plant cost
    - Indirect cost
    - Fixed capital
- Operational cost in the revised situation
- Quantification of sustainability improvement using a suitable Key Performance Indicator

---

At this point I need to make one thing very clear. I assume that your company has its own Process Design and Engineering people capable of developing the deliverables in Table 12. Experience has taught me that every company has its own way of setting these things up. I therefore don't intend to interfere with your working routines, let alone telling you how it should be done.

However, from practical experience I have learned integration of a new piece of equipment into an existing plant deserves special attention. I like to share my learning points with you. This is what Section 4.2 is about.

4.2 Sustainable design guidelines

You are looking at a new piece of equipment that is going to replace one or more Process Functions in your existing plant. This will affect each of the items in Table 12. Naturally, I am unaware about the specifics. From practice I know it can be a real challenge to integrate new equipment in an existing plant in a sustainable way. More often than not I have seen "shortcuts." A shortcut is a fix to get the new kit quickly up and running in the plant. These fixes often use old school methods like deployment of fossil fuels or hardcore chemicals like sulfuric acid, hydrochloric acid or caustic soda.

From a Team point of view this is understandable. Your Team members have put a lot of effort into identifying a sustainable solution. They found one. They want to put it to work asap! However, applying a quick fix may neutralize or even counteract the sustainability gain you are pursuing. Empirics has yielded a set of guidelines that enhance the chances of a sustainable integration. I have discussed each one of them and have summarized in Table 13. The discussion order is arbitrary.

Obviously, if you are trying to improve the reactor part of the process, the best thing you can do is to create a product yield of 100%. No byproducts. No recycling. No energy required for the separation of byproducts from main product. Essentially no real "separation backend" of your process, which reduces capex. The separationists can go home. This may sound silly or faint in the ears of experienced chemical engineers who know that most conversions per pass end up well below 100%. Still, setting a yield target of 100% focuses your mind. It helps you in looking for ways to get as close as possible. If you are looking at equilibrium reactions, continuously take away one of the components on the product side of the equation to push the reaction to completion (le Chatelier). This can be done by physical separation (extraction, absorption, adsorption, membrane separation, complexation, etc.) or by chemically reacting away one of the components on the product side. If you are looking at other nonequilibrium reactions: first look at and fully understand the kinetics. Then design the equipment around the chemistry that best supports these kinetics. Again this may seem trivial, but in many existing manufacturing processes it is the other way around. Most of the time, the chemistry is adapted to match the equipment in which the conversion has to take place.

Next we have to discuss our endless love for solvents. It seems to be the first thing an organic chemist does in his lab: dissolving starting materials in a solvent to get a reaction going. Much later, the chemical engineer inherits the newly found process from the organic chemist to scale it up. That is exactly what he does: he scales it up. He does not ask a critical question like: "Why do you use solvent?" Apparently, the use of solvents has become something we professionals do not question or discuss. This sounds much like the "copied behavior" we described in Section 2.4. Unless your process manufactures a solvent, the solvent used is not the product. At some stage separation of product and solvent is required which costs energy and makes your process less sustainable. Check whether you can use an intermediate or the product itself as a solvent. If this is not possible, use as little solvent as the process permits. In a broader sense avoid the use of "helping" chemicals. These may be helpful at some stage in the process but they may hit you downstream, causing fouling and/or separation issues.

Then there is our seemingly insatiable desire to evaporate. For front-end conversion sections there may be good arguments to execute a reaction in the gas phase like improved reaction rate, improved selectivity or higher mass transfer. Here the question is: can it also be done in liquid phase? For backend separation sections again I often see "copied behavior." For decades now, people almost automatically go for distillation.

In the early days of personal computing the first simulation tools hitting the market were distillation programs. I was often led to believe, unprovably, that this fact promoted the interest in distillation. When your only tool is a hammer, every problem looks like a nail. In the Netherlands, some 65% of the energy used by the chemical sector is used for distilling product mixtures! Most of this energy comes from burning natural gas. Only a minor part of the required heat comes from waste heat. My educated guess is that this is true for most European countries with a significant chemical industry. I am not sure how this plays out on other continents.

A simple argument for keeping everything in liquid phase is vaporization costs energy. When you operate a process with a structurally high heat demand (which follows directly from the Process Flow Diagram your company has chosen) you are in for a treat when energy prices go up. In this situation cost-cutting and CO_2-emission reduction go hand in hand. Also, if vaporization temperature approaches degradation temperature, product quality goes down. Anyone familiar with operating a distillation section in a chemical plant knows what disastrous things a high reboiler temperature can do to a liquid made up of beautifully structured yet temperature-sensitive molecules. And finally, vapor-phase operation boosts equipment size thus capex and space requirements. The message is that if it is possible in liquid phase, don't go for gas phase. It enhances process sustainability and reduces capex and opex at the same time.

Closely related to this subject of evaporation is the next item: water. We like water. We know water. We don't mind water in our process. The point is this. Since water has the highest heat of vaporization of all substances known to men, it immediately hits you once you start thermal unit operations like drying or dewatering. Water is not our friend in more than one sense. Please go for mechanical dewatering, e.g., using force fields (centrifugation) or pressure gradients (membrane processes). If water is a reactant, prevent evaporation. If water is a solvent, try and find another solvent.

In everyday life we often have to deal with impurities. Mostly, these are unwanted byproducts generated in the reaction section of the plant. Right at the beginning my statement is to get rid of these as early in the process as you can. Impurities distributing themselves downstream in your process will be a continuing source of trouble: blockages, fouling, loss of heat transfer, clogging, off spec product, etc. With distribution comes dilution. As impurities get more downstream, the lower their concentration is. The more difficult it becomes to get them out of your plant. My advice is to focus on the process stream with the highest impurity concentration and remove the impurities by physical or chemical means. Try to accomplish this on one specific spot. This spot you can design in such a way you deliberately cause trouble, e.g., by forcing the impurities to precipitate. Make sure that you can block-in this part of the process. This way you can continually clean the spot without a total shutdown of the plant.

Reactant concentration is another handle on your process. When reactions are very exothermic and/or products are dangerous in terms of explosive or toxic, people tend to dilute with solvent. In addition to this they lower reaction temperature. Both dilution and temperature drop cause reaction rates to go down significantly. We end up with large reactor inventories containing lots of solvent and very low production rates. Needless to say that this impacts safety and capex negatively. My advice is to try and go completely the other way. Explore the limits of high concentration of reactants that enable fast conversion rates without undesired side effects. Maintain temperature on a level where reaction is fast, push the concentrations of reactants to the limit explored and execute the reaction in a reactor that can deal efficiently with high heat loads. That way you have small inventories, high production rates and small equipment. Good for safety, process economics and capex. Admittedly, this goes against the grain but it works. The Holy Grail here is the reactor that manages high heat loads. These reactor types exist and they can be bought off the shelf. Not batch, not large.

Next to reactant concentration is reactant ratio. You may visualize one-on-one reactions like A + B yields C. This may have a few intricacies, depending on the number of liquid/gas phases involved, which reactant is in which phase, etc. In the case of a one-phase system with two components and one solvent, both component concentrations will play a role in kinetics. You may argue excess of one-component boosts reaction rate. This may be observed in practice. On the other hand, you end up with a product mixture containing one component in excess. The separation issue will be knocking on the door. Operating with equimolar quantities of reactants in this specific case may be the best way forward at the expense of reaction rate. I don't see a general rule of thumb concerning advised reactant ratio's for all possible reaction systems.

Then there is recycling. In the sense of reusing waste materials recycling is a route worth pursuing. In the sense of sending part of a process flow upstream again because you cannot get it right the first time, it is a red flag. A recycle boosts the size of many equipment items. If there are temperature swings on the route back, a recycle increases energy and/or cooling water consumption. If it also involves liquid/gas-phase changes and its associated pressure drop and heat of vaporization, required compressor power and additional heat generation drive energy consumption to an even higher level. And in case you have not dealt yet with impurities, these are more than happy to join the recycle. The more the merrier. Accumulation of dirt is your reward. This issue can be tough to solve, unless you can turn the issue into an opportunity (see "Impurities" mentioned above). To summarize that recycles most of the time are indicators of incomplete conversion or poor separation. They cause equipment and piping involved in the recycle to become larger. Capex goes up. Opex goes up as well since recycling requires additional heat and power. Thus sustainability goes down.

Gravity (Table 8) is often overlooked when we want to transport solids or liquids inside the plant from A to B. It is available everywhere and it is for free. When you want to consider this I recommend to analyze your process first. What are your liquid viscosities? What is the cumulative friction force of ducts that has to be overcome for transporting liquids? What is the required speed of transport of (s/l) materials? If all answers look positive you can give it serious consideration.

I wrap up with the example touched upon at the beginning of this paragraph. Avoid the use of old school neutralizing agents like hydrochloric acid or caustic soda. Solutions of HCl and NaOH in water can quickly solve the issue of an undesired high or low pH in your process. However, this is at the expense of producing large amounts of salt. This is not sustainable. If pH adjustment is required, consider integrating regeneration units that restore the original acid/caustic solutions for reuse (e.g., membrane systems). Table 13 summarizes the Sustainable Design Guidelines during Plant Revamp.

Table 13: Sustainable Design Guidelines during Plant Revamp.

	Design guideline	Arguments	What you can do
1	Boost product yield per pass in the conversion stage to the maximum.	A yield of 100% renders the separation backend of your plant obsolete.	Equilibrium reactions: continuously take away one of the components on the product side of the equation to push the reaction to completion; this can be done by physical separation (extraction, absorption, adsorption, membrane separation, etc.) or by chemical reaction. Other reactions: first look at and understand the kinetics; then design equipment around the chemistry that supports these kinetics.
2	Use a minimum set of solvents.	Solvents are not the product. At some stage separation is required which costs energy and makes your process less sustainable.	Use intermediates or the product itself as a solvent.
3	Avoid introducing other solvents or "helping" chemicals.	id	id
4	Keep all reactants and products in the liquid phase to avoid vaporization.	Vaporization costs energy. If vaporization temperature approaches the degradation temperature, product quality goes down. Vapor-phase operation boosts equipment size, thus capex.	Reactions: try and execute them in liquid phase. Separations: avoid evaporation.
5	Avoid the use of water when there is a chance it evaporates.	Water has the highest heat of vaporization of all substances.	If water is a reactant: prevent evaporation; if water is a solvent, try and find another solvent.

Table 13 (continued)

	Design guideline	Arguments	What you can do
6	Remove impurities from the product mixture as early in the process as possible.	Impurities distributing themselves downstream in your process can be a source of trouble: blockages, fouling, loss of heat transfer, clogging, off spec product, etc.	Focus on the process stream with the highest impurity concentration and remove them by physical or chemical means.
7	Work with concentrated streams as much as possible.	Generally, reaction rates go up with increasing concentration of reactants; volumes of process units become smaller.	Explore the limits of concentration of reactants that enable fast conversion rates without undesired side effects.
8	Avoid the use of (additional) water especially when evaporation is in play.	Water is among the substances with the highest heat of vaporization. Evaporation of water directly translates into higher energy use and a less sustainable process.	Select alternative components that have the same physical/ chemical function as water without the associated high heat of vaporization.
9	If impurities do go along with the product downstream try and catch them in one controlled spot.	If you don't catch them on one spot, you will have to catch them everywhere they go.	Cause deliberate process disturbance that affects impurities (e.g., sudden phase change, temperature/pressure swing, reaction, degradation and precipitation).
10	Interpret the existence of recycles as a red flag: take a step back and look for ways to remove the recycle.	Recycling most of the time is an indicator of incomplete conversion or poor separation. They cause all equipment involved in the recycle to become larger. Capex goes up. Opex goes up as well since recycling requires additional heat and power. Thus sustainability goes down.	Look at the reaction yield and/or the quality of the separation step involved. Try and improve these to a level where recycling can be omitted.
11	Operate with equimolar quantities of reactants and avoid excess of one component.	This holds for one-on-one reactions in the same phase; there is no general rule.	

Table 13 (continued)

Design guideline	Arguments	What you can do
12 Use gravity for transporting solids and/or liquids in your plant whenever you can.	Gravity is available everywhere. It is for free. It is sustainable.	Analyze whether your process allows for transport by gravity, e.g., by looking at liquid viscosities, required transport speed of (s/l) material, cumulative friction force of ducts that has to be overcome.
13 Avoid the use of old school neutralizing agents (HCl, NaOH).	Solutions of HCl and NaOH in water can quickly solve the issue of high/low pH in your process but at the expense of producing large amounts of salt which is not sustainable.	If pH adjustment is required consider integrating regeneration units that restore the original acid/caustic solutions for reuse.

Note: In addition to these guidelines I like to cite the following heuristics from other authors [48, p. 113; 49]:
Heuristic A: Choose the reaction system with the highest selectivity to the product; this looks like guideline [1] but it recognizes the fact that 100% conversion is hard to achieve.
Heuristic B: Do not consume additional chemicals for reactions or separations; this is similar to [3].
Heuristic C: Do not use stripping with nitrogen or washing with water; this nitrogen stripping advice is new; the advice on washing with water resembles [8].

4.3 Consequences of circular innovation and deployment

Making a choice for circular innovation and deployment will change things inside your company. From an organizational point of view the introduction of circular innovation has consequences. Every 2–3 years, there will be a new innovation cycle that needs proper preparation, execution and followup as has been described in Chapter 3. This is a new recurring work process that has to be implemented. A certain amount of manpower needs to be reserved.

Financially, the choice for circular innovation and deployment may have the largest impact. There will be direct and indirect effects in this respect. Direct effects start with relatively moderate in kind costs of every employee involved in the new work process and cash out costs of 2-day Workshops. They end with Investment Decisions for deployment of the newly found equipment in the existing plant. Depending on the size of the revamp project the Investment Decision may also comprise Risk Assessment, Legal and Regulatory Considerations, Project Implementation and Financial Performance [63]. Each of these elements is a cost factor. Circular innovation leads to a higher average investment level. Assuming that profits will be lower due to more downtime, the annual dividend for shareholders will drop. This is the indirect effect.

On the plus side the company's manufacturing process gets more and more sustainable. Also here there are direct and indirect effects. A direct effect is that CO_2 emissions will go down, which is a financial plus. An indirect effect is that the company keeps pace with its competitors by participating in a repetitive sustainability improvement effort. Being on par with fellow companies will be seen and judged by the market. This will affect what customers are willing to pay for your product.

What does this all boil down to? In the short run, the company is looking at new working routines and an additional workload which may be partly mitigated by reorganizing existing work processes. After each new deployment management has to deal with higher operational risks for a certain period of time which will lead to more downtime of the plant. This is the consequence of getting to know and operate a new piece of equipment in the existing plant. The shareholders are looking at lower returns on their shares and they have to live with a company that is on average more risk-taking than before. Repetitive step-ups in process sustainability are the result.

In the long run the continual increase of process sustainability will have impact. The company will stay in the game with competitors who are also improving on this topic. Another effect may be this effort may enhance the long-term continuity of the company. From consultancy practice I know that this is particularly true for family-owned chemical businesses. For example, management of one such company was explicit about this. They wanted to hand over control to the next generation with a state-of-the-art manufacturing process including proper sustainability performance to make it future proof. You can read a specific value set between these lines. This goes beyond making a decent profit. It is also about caring for the next generation and about making sure that the company can withstand future challenges successfully. Resilience may be the adequate term here (Explainer 1).

The choice for circular innovation is of course entirely up to the company. In essence we are talking about the value sets we discussed earlier in Sections 1.6 and 1.7. My role is merely to point this out and make you aware.

4.4 Summary

It is essential to plan and execute the development of a PRP directly after the 6-steps SIM Workshop to keep your Support Base alive and to keep momentum going. More often than not valuable sustainable solutions emerging from the Workshop are integrated in the plant by means of a quick fix, thus lowering or even counteracting the pursued sustainability increase of the process. From practical experience, 13 Sustainable Design Guidelines are discussed, each of which has the purpose of a sustainable integration. Introducing circular innovation in the short run affects the working processes. It also has financial implications. In the long run it may ensure company continuity and enhance company resilience.

5 Sources of inspiration

Abstract: This chapter covers three sources of inspiration. First, there is an extensive body of 81 intensified chemical technologies. The history and impact of intensification thinking on design and engineering are briefly recapitulated. In order to make the overview of 81 technologies usable during the 6-steps SIM Workshop, each of these is characterized in terms of process function, driving force applied and maturity level. Second, there are renowned international exhibitions like ACHEMA, which has an inspiring website. Analysis shows both innovation and sustainability are becoming mainstream. Third, there is the set of offerings concerning innovation and sustainability from the worldwide top 20 process engineering companies. About 95% of these companies are explicitly offering services and/or equipment that is innovative and/or sustainable. The deadlock of "no demand for sustainable – no design for sustainable" that has lasted for decades is gone.

This content is meant for use during the 2-day SIM Workshop. Trial and error have taught me it is most effective when you present the content of Appendix 2 as an Excel sheet on easy to use laptops on tables near seats and sofas in the Workshop room and in adjacent rooms. Please mention the existence of these laptops a few times in a low-profile manner. This makes the threshold low for workshop participants who take a sneak preview during a coffee break.

5.1 History of intensified chemical technologies

When process intensification (PI) kicked off in the UK, Germany and the Netherlands somewhere between 1994 and 1998, it was regarded by industry as something radically new. Nowadays, most of the new technologies that are c.q. have been developed to commercial scale are intensified. The idea of intensifying appeared to be very inspiring for technological communities in academia and research institutes. It sparked the founding of dedicated knowledge networks and technology development programs in several countries in Europe and in the USA. Nice examples comprise the RAPID Institute in the USA [50], the Working party on PI of the European Federation of Chemical Engineering (EFCE) [51], Europic [52]: a European Network of Companies on Process Intensification, SPIN-NL: the Dutch Knowledge Network for Sustainable Process Intensification and PIN-UK: the English Process Intensification Knowledge Network. A comprehensive overview of the "Werdegang" of PI has been written by Frerich Keil [53]. The European Roadmap of PI covers both Methods and Equipment [54].

I use two working definitions of PI:

– A process is intensified when two or more unit operations are integrated into one shell. Examples: reaction and distillation, reaction and membrane separation, mixing and reaction, reaction and extraction, etc.

https://doi.org/10.1515/9783111383668-005

– A piece of equipment is intensified when it is up to 10 times smaller in physical size, footprint and/or weight and it still delivers the same performance as the former non-intensified piece of equipment in terms of output or transfer of heat, mass or impulse.

Examples: replacing a large evaporating batch reactor enabling a highly exothermic reaction with hours of residence time with a small spinning disk reactor with minutes of residence time; replacing distillation with membrane separation; omitting solvent by using an intermediate as a dissolving agent; etc.

Figure 20 shows a somewhat different breakdown of the concept of PI. It was published for the first time in the European radmap of Process Intensification.

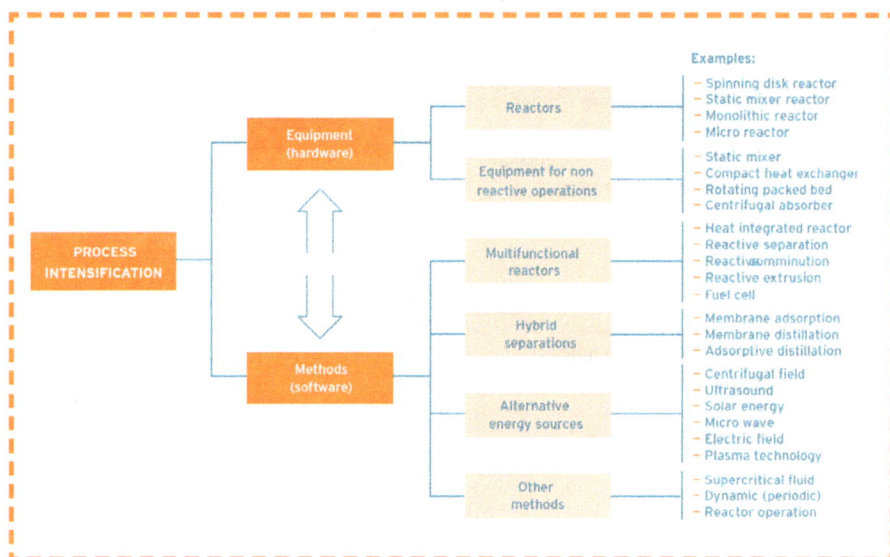

Figure 20: Elements of process intensification.
©The European Roadmap on Process Intensification

The European Roadmap for PI [54] made a start with categorizing and listing all known PI technologies. Since 2007 this list has grown in my consulting practice from 72 to 81. In order to make the list usable during the 2-day Workshop I have added process functions, driving forces and maturity level to characterize each of the listed technologies.

A brief overview is given in Table 14. Appendix 2 offers an extended overview.

5.2 Overview of intensified chemical technologies

Process function, driving force and maturity level of each of these technologies can be found in Appendix 2. Giving further details about these technologies is beyond the

Table 14: Overview of intensified chemical technologies.

1	Adsorptive distillation	42	Microwave reactors – polymer
2	Adv. HEX – plate	43	Microwave separation
3	Adv. HEX – spiral	44	Millisecond reactors
4	Centrifugal adsorption technology	45	Molecular-imprinted polymers
5	Centrifugal extractors	46	Monolith reactors
6	Chemical looping	47	MPPE
7	Continuous oscillatory baffled reactors	48	Multistream heat exchangers
8	Cryogenic separations	49	Other structured catalytic reactors (KATAPAK's, parallel passage, etc.)
9	Distillation-pervaporation systems	50	Pervaporation-assisted reactive distillation
10	Dividing wall columns	51	Photochemical reactors
11	Ejector (Venturi)-based reactors	52	Plasma reactors
12	Electric field-enhanced mixing	53	Pulsed chromatographic reactors
13	Electric field-enhanced operations – other (e.g., fouling prevention)	54	Pulsed combustion drying
14	Electric field-enhanced extraction-dispersion	55	Pulsed compression reactor
15	Electric field-enhanced heat transfer	56	Pulsing operation of multiphase reactors
16	Electrochemical reactors	57	Reactive absorption
17	Extractive crystallization	58	Reactive comminution
18	Extractive distillation	59	Reactive condensation
19	Foam reactors	60	Reactive crystallization/precipitation
20	Gas-solid-solid trickle flow reactor	61	Reactive distillation
21	Heat-integrated distillation columns	62	Reactive extraction
22	Heat-integrated distillation	63	Reactive extrusion
23	Hex reactor	64	Reverse flow reactors
24	Hydrodynamic cavitation reactors	65	Rotating annular chromatographic reactor
25	Impinging streams reactor	66	Rotating packed beds
26	Induction – ohmic heating	67	Rotor-stator mixers
27	Ionic liquids	68	Simulated moving bed reactor
28	Membrane absorption/stripping	69	Sonochemical reactors
29	Membrane adsorption	70	Spinning disk reactors (SDRs)
30	Membrane crystallization	71	Static mixers
31	Membrane distillation	72	Static mixers-heat exchangers
32	Membrane extraction	73	Static mixers-reactors
33	Membrane reactor (selective)	74	Structured internals for mass transfer operations
34	Membrane reactors (nonselective)	75	Supercritical reactions
35	Micro(channel) reactors	76	Supercritical separation
36	Micromixers	77	Supersonic gas-liquid reactors
37	Microchannel heat exchangers	78	Supersonic gas-solid reactors
38	Microwave drying	79	Ultrasound-enhanced crystallization
39	Microwave heating	80	Ultrasound-enhanced phase dispersion/mass transfer
40	Microwave reactors – cat	81	Viscous heating
41	Microwave reactors – non-cat		

scope of this book. Detailed descriptions of various PI technologies are given by Harmsen and Verkerk [48] and Stankiewicz et al. [49]. Further information about most technologies can be found in the public domain.

5.3 Cost estimating intensified technologies

Once you embark on cost estimating one of the technologies listed in Table 14 for a specific deployment, you will stumble over the limited availability of cost data. This is particularly true for a country like the Netherlands. The DACE – the Dutch Network and Knowledge Center for Cost Engineering and Value Management – publishes a *Price Booklet* annually [64]. DACE offers cost data on most pieces of standard process equipment. The booklet is quite handy and has become part of every cost estimator's library in the Netherlands.

However, cost data of most technologies listed in Table 14 are absent. This means, you have to set up close cooperation between your process design engineer and your cost estimator. Together they have to generate cost data by dividing the technology at hand into parts, designing each part from the ground up using known sources, producing subestimates and compiling these subestimates in an overall estimate. Clearly this is a labor-intensive exercise with a reasonably large uncertainty.

This being the Dutch situation one may ask on a general level how likely it is that companies will come up with Plant Revamp Proposals containing intensified technology when cost estimators cannot produce a cost estimate on their own due to lack of data. This sounds like a chicken-and-egg situation. What is the point of proposing PI technology if your cost estimator cannot quantify the investment cost? Revamp Proposals may have to stick to old school technology. Intensified designs do not get built and deployed. Therefore, no new cost data for cost estimators are generated. How to resolve this situation? For several years, Traxxys has participated in the Working Party on cost estimating of DACE. The objective is to generate cost data of intensified technologies and integrate these into the *Price Booklet*. I have no information about the availability of cost data of PI technologies in other countries. I would not be surprised if this situation also occurs elsewhere.

5.4 How to use the overview

Appendix 2 gives an overview of intensified chemical technologies. Each of the pages in this appendix is taken from the original Excel overview depicted in Figure 21.

The tables in Appendix 2 extracted from Figure 21 are to be used rotated 90° clockwise. On the left, you will notice a column with 81 different intensified chemical technologies in alphabetical order. At the top you see three types of headers from left to right. All items in the headers are in alphabetical order. The first header comprises process functions. You may have noticed these functions earlier in Table 4. The second header

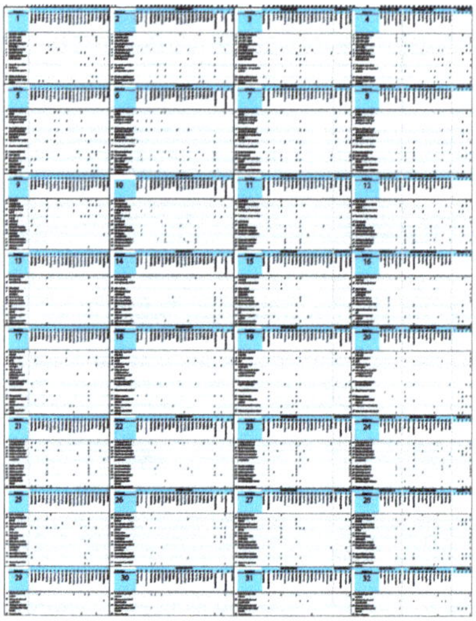

Figure 21: Global overview of intensified chemical technologies in Excel.

lists driving forces. They correspond to the driving forces given in Table 5. The third header lists the maturity level of each technology expressed as TRL. A full characterization of each technology is given in a row covering four tables. This is why the overview in Table 21 is repeated in Appendix 2 in clusters of four tables: 1–4, 5–8, 9–12, etc.

A first important remark is that this is a living table. By this I mean the number of technologies listed, the types of process functions, the number of driving forces and the maturity levels are all time dependent. The table can best be characterized as a snapshot. When we take the next snapshot 2 years from now, the table will look different.

When can Appendix 2 best be used? Figure 19 answers this question. The appendix can help you in step 2 where you identify the process function of the equipment that causes a bottleneck in the process. You can use it to identify the underlying driving force in step 3. It gets really interesting when you want to identify other driving forces that can drive the same process function in step 4. Next, in step 5, it can help you in translating the process function driven by a new driving force back into tangible equipment. Finally, it can help you in step 6 in classifying the list of newly found pieces of equipment into three categories of TRL level.

A second important remark is this: Figure 19 and Appendix 2 are merely tools to structure and direct the creativity you have sparked in your Team. Appendix 2 is not meant as a border you should not cross in the sense that if a newly found option is not in the table it is "wrong." It may be that a Team member comes up with a new driving force to drive the process function, thus solving the bottleneck issue. To subsequently discover in step 5, there is no piece of equipment with this specific [driving force]/[process function] combination present in Appendix 2. If this occurs, it may be that Appendix 2 is incomplete (i.e., there is equipment on the market with this [driving force]/[process function] combination) or that your Team member has found a lead for development of a new type of equipment not yet on the market.

Note. You can download the latest version of this overview of intensified chemical technologies (Appendix 2 in this book) in Excel format from de Gruyter's website using the password that comes with a hard copy or an e-copy of the book. The Excel version contains active filters that enable you to make quick selections based on process function, driving force or maturity level [55]. Obviously, the user-friendliness of the Excel tool on a laptop/PC is better than the hard copy or e-copy version.

5.5 International exhibitions and process engineering contractors

For decades, you were caught between a rock and a hard place when you asked an engineering contractor if his/her technology portfolio contained innovative and/or sustainable technologies or services. The answer was invariably the same, whichever company you asked: "We could build this, but our clients don't ask for it." Ok! Let's go then to their clients and ask them the same question. Same question – same answer: "We could build this, but our engineering contractors don't build it." A circular deadlock of "no demand for sustainable – no design for sustainable."

Slowly and quietly, this chicken-and-egg situation has changed. More and more engineering contractors have sustainable technologies in their portfolio. I present two ways to give you a feel for the relevance of innovation and sustainability. The first way is a brief analysis of all items presented at the greatest exhibition in the world for the process industries. ACHEMA in Frankfurt (Germany) opens its doors triannually. It covers products, technologies, congress presentations, public press articles and ACHEMA Magazine articles. The second way is to create a list of the top 20 process engineering companies and screen their websites for products and services related to innovation and sustainability. An overview of relevant ACHEMA data is shown in Figure 22.

At ACHEMA 2024, innovation scores seven times higher than sustainability in exhibitor offerings. Combining "Exhibitors" and "Products" data, the type of innovation must be services rather than tangible hardware. Sustainability scores three to four times higher than innovation in all other interest groups. In absolute terms, sustainability is present in 9–37% of items presented and/or discussed. It scores highest in the ACHEMA congress, which was organized in parallel to the exhibition. In conclusion, we can state both innovation and sustainability are becoming mainstream in the world's greatest event for the process industry. A photo impression underlining and confirming this conclusion can be found inside the front and back covers of this book. The ACHEMA website offers many inspiring examples of innovative and sustainable technologies [59].

Having looked at the ACHEMA exhibition, we now go to a group of 21 leading engineering companies to see whether their offerings include innovation and/or sustainability (Table 15). A green bar implies the subject is explicitly covered on the website of the company.

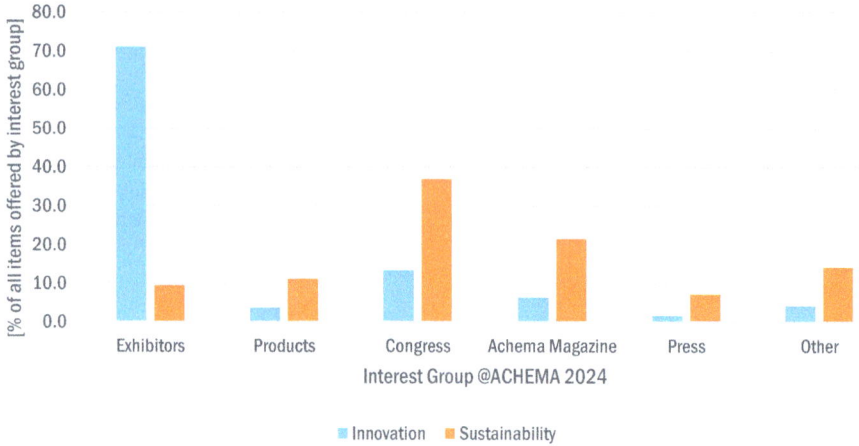

Figure 22: Occurrence of innovation and sustainability offerings at ACHEMA 2024 Frankfurt [59].
Note: Data have been taken from the ACHEMA website after the 2024 event was closed.

Table 15: Top 21 process engineering companies [56, 57, 60, 61].

	Engineering company	Website	Innovation	Sustainability
1	ABB Lummus Global	http://www.abb.com/		
2	Aker Solutions	https://www.akersolutions.com/		
3	Bechtel	https://www.bechtel.com/		
4	Fluor Corporation	https://www.fluor.com/		
5	General Electric Vernova	https://www.gevernova.com/		
6	Hitachi	https://www.hitachi.com/		
7	Honeywell UOP	https://uop.honeywell.com/		
8	IFPEN	https://www.ifpenergiesnouvelles.com/		
9	Jacobs	https://www.jacobs.com/		
10	Johnson Matthey	https://matthey.com/		
11	Larson & Toubro	https://www.larsentoubro.com/		
12	Quanta Services	https://quantaservices.com/		
13	Schlumberger	https://www.slb.com/		
14	Siemens	https://www.siemens.com/		
15	Spirax-Sarco	https://www.spiraxsarco.com/		
16	Technip	https://www.technipfmc.com/		
17	Tetra Tech	https://www.tetratech.com/		
18	Thyssenkrupp	https://www.thyssenkrupp.com/		
19	Topsoe	https://www.topsoe.com/		
20	Uhde	https://www.thyssenkrupp-uhde.com/		
21	Worley	https://www.worley.com/		

This list I have put together by combing information about top engineering companies and screening their websites for content about innovation and sustainability on the process technology and energy webpages. Interestingly, the majority of these companies put more emphasis on sustainability than on innovation in their offerings.

As a spinoff of this exercise, I recommend you to take a look at the quoted websites, many of which are very inspiring to read.

Combining Figure 21 and Table 10, we can conclude that a top-tier exhibition event like ACHEMA emphasizes sustainability in the same manner renowned engineering companies do.

Apparently, both innovation and sustainability have entered the mainstream market of equipment and services for the process industries. Chemical companies looking for innovative and sustainable solutions on the market are no longer caught between a rock and a hard place.

5.6 Summary

Chapter 5 offers three sources of inspiration that may be instrumental during the 2-day Workshop.

The first source encompasses an extensive list of 81 intensified process technologies, each of which is characterized in terms of process function, driving force applied and maturity level. When installed on a laptop, selection filters can be applied that enable quick subselections at the will of the user. This tool will be effective in various stages of the 6-steps SIM.

Lack of cost estimating data for intensified technologies may hamper your initiative to develop Plant Revamp Proposals, including one such technology. This issue needs solving to lower the threshold of deployment of PI technologies in the industry at large.

The second source is the body of international exhibitions organized for the process industry. ACHEMA is taken as an example. A brief analysis shows that both innovation and sustainability are becoming mainstream.

The third source is a list of 21 top-tier process engineering companies with hyperlinks to their respective websites. Most of these companies show innovative and/or sustainable technologies or services in their offerings that may be applicable during the 2-day Workshop as well. The advantage of these offerings is that they are market-ready.

6 Epilogue

Let's first look at the answer to the question in the preface: how can it be so many sustainable technological solutions already exist and so little companies in the process industries actually deploy them?

The answer is simple. There are no internal drivers for a successful company to go for innovation or sustainability. "Successful" as in the present-day definition of the game: key economical parameters like EBITDA, ROI, RTEP and NPV are all in the right range. Change would endanger this. Shareholders, managers, employees and family; nobody wants this to happen. However, history has shown that external drivers may emerge that can structurally affect the course of any company. Examples comprise Total Quality Management, Health, Safety and Environment, Compliance, Sustainability Benchmarking and – most recently in Europe – the Corporate Sustainability Reporting Directive.

Let's then look at the journey you have made to arrive at this point.

First, you will have executed a situational analysis of your company. Second, the guide has given you practical definitions of innovation and sustainability you can work with. Third, it has described and explained a proven method that enables you to innovate your process yourself toward a higher level of sustainability. And fourth, it has given you a set of Sustainable Design Guidelines to make sure that you integrate your innovation in a sustainable way.

What deliverables did you get? Based on the outcome of aforementioned analysis, the guide will have advised you on a course of action, keeping in mind that added value to the company is the first criterion. This implies that you will have defined a sustainability project that adds value and is in line with the real purpose of the company. A project that may anticipate, mitigate and manage the impact of emerging external drivers on your company. You will have built a Company Support Base with colleagues that support you in realizing your initiative. You will have formed an enthusiastic Company Innovation Team that has the right mix of mindset and skills to take on the innovation challenge. You will have secured budget for two out of four phases of your total project (Workshop and Design to Deploy). The guide will have provided you with workable definitions of innovation and sustainability. Armed with these features you and your Team will have executed the 2-day Workshop applying 6-steps SIM. The Workshop will have yielded a number of innovative types of equipment. Directly after, in the Design–to-Deploy phase, you will have quantified the equipment identified as most promising into a Plant Revamp Proposal, using Sustainable Design Guidelines. This PRP will be ranked together with other company investment proposals following company procedure. Let's hope that your proposal meets investment criteria!

My guess is that you will have acquired an appreciation for what innovation toward sustainability of existing processes entails. On company level your initiative re-

https://doi.org/10.1515/9783111383668-006

quires congruence with company objectives. On colleague level it requires insight in people's motivation as well as convincing skills. On personal level it requires courage, perseverance and the ability to operate outside your comfort zone and to maintain open communication lines with key people in your company.

Assuming that you and your Team have conquered all the hurdles and the company ranking committee has honored your Plant Revamp Proposal with budget, I like to congratulate you all with this sustainable result!

References

[1]   Janka Stoker; Harry Garretsen;"Goede leiders in onzekere tijden – lessen voor organisaties en de politiek"; page 99; Business Contact, 2^e druk juni 2023, ISBN 978 90 470 1676 2; Amsterdam, Antwerpen

[2]   Henry Mintzberg; "Structure in Fives: Designing Effective Organizations"; Prentice Hall International Editions; ISBN 0-13-854191-4

[3]   History of Total Quality Management: https://asq.org/quality-resources/total-quality-management/tqm-history; website visited on 15-4-24

[4]   HSB Solomon Associates LLC: https://www.solomoninsight.com/about/; website visited on 11-4-24

[5]   3d: https://www.solomoninsight.com/services/benchmarking/; website visited 10-4-24

[6]   Solomon on Sustainability: https://www.solomoninsight.com/enterprise/capital-portfolio/sustainability-investment-analysis; website visited 11-4-24

[7]   53 Different Topics on Sustainability: https://www.solomoninsight.com/enterprise/sustainability-strategic-insight/sustainability-performance-analysis; website visited on 10-4-24

[8]   McKinsey on Sustainability: https://www.mckinsey.com/capabilities/sustainability/how-we-help-clients; website visited 15-4-24

[9]   IBM on Sustainability: https://www.ibm.com/sustainability?utm_content=SRCWW&p1=Search&p4=43700077628005548&p5=p&gad_source=1&gclid=CjwKCAjwoPOwBhAeEiwAJuXRhwD8ZKvz_Hl1aQLAFyxwEQ8UJYC0k42g58TSE3AT02XJejHgv7k3bhoCFEkQAvD_BwE&gclsrc=aw.ds; website visited on 15-4-24

[10] Deloitte on Sustainability: https://www.deloitte.com/global/en/issues/climate/sustainability-and-climate.html; website visited on 15-4-24

[11] Ernst & Young on Sustainability: https://www.ey.com/en_gl/services/sustainability/ey-center-for-sustainable-supply-chains; website visited on 15-4-24

[12] KPMG on Sustainability: https://kpmg.com/xx/en/home/about/our-impact-plan/planet.html

[13] PwC on Sustainability: https://www.pwc.com/gx/en/services/sustainability.html; website visited on 15-4-24

[14] MVO on CSDR: https://www.mvonederland.nl/wat-is-de-csrd-richtlijn-en-hoe-ga-je-ermee-aan-de-slag?gad_source=1&gclid=Cj0KCQjwncWvBhD_ARIsAEb2HW_hEFO3dtCqDef-R4V-fLe1AcL-PhYqzxONdG752Vp-eSaVtJoaW-AaAiKyEALw_wcB; visited on March 27, 2024

[15] Management Fads and Fashions: Leading and Directing Companies by Popular Codified Methods; Carl V. Rabstejnek, P.E., M.B.A., Ph.D. www.HOUD.info/Fads.pdf; website visited 4-4-2024

[16] Abjihit Banerjee; Esther Duflo; "Good Economics for Hard Times"; 2019; Penguin Books limited; ISBN 9780241306895

[17] Stephani Kelton; "The Deficit Myth – How to Build a Better Economy"; 2021; Publisher Hachette Collections; ISBN 81529352566

[18] Kate Raworth; "Doughnut Economics"; 2018; Cornerstone Publishers; ISBN 9781847941398

[19] Paul Schenderling; "Er is leven na de groei"; 2022; Bot Publishers; ISBN 9789083256443

[20] Mariana Mazzucato; "The Entrepreneurial State"; 2018; Penguin Group Publishers; ISBN 9780141986104

[21] Carlotta Perez; "Technological Revolutions and Financial Capital"; 2003; Edward Elgar Publishers; ISBN1843763311

[22] J.K. Simmons as CIA Superior – Burn After Reading (2008), a movie by the Coen Brothers: "What did we learn, Palmer?"

[23] Ben van Beurden, former CEO of Shell; "Non Solus: New Energy for the Netherlands (and the world) with a Translation in English"; EW books; 2018

[24] Bernard Burnes; "The origins of Lewin's three-step model of change"; The Journal of Applied Behavioral Science; 56(1); 32–59; 2020; © The Author(s) 2019

https://doi.org/10.1515/9783111383668-007

[25] Jan Rotmans; "Transitiemanagement – sleutel voor een duurzame samenleving"; 2003; Koninklijke van Gorcum B.V.; ISBN 90-232-3994-6

[26] Niklas K. Steffens; S. Alexander Haslam, Stephen D. Reicher; Michael J. Platow; Katrien Fransen; Jie Yang; Michelle K. Ryan; Jolanda Jetten; Kim Peters; Filip Boen; Leadership as social identity management: Introducing the Identity Leadership Inventory (ILI) to assess and validate a four-dimensional model Niklas; The Leadership Quarterly; 25; 1001–1024; 2014

[27] Rolf van Dick; Jérémy E. Wilson-Lemoine; Niklas K. Steffens; S. Alexander Haslam; Identity leadership going global: Validation of the Identity Leadership Inventory (ILI) across 20 countries; Journal of Occupational and Organizational Psychology; 91(4); 696–728; 2018

[28] John Antonakis; Nicolas Bastardoz; Philippe Jacquart; Boas Shamir; "Charisma: An ill-defined and ill-measured gift"; Annual Review of Organizational Psychology and Organizational Behavior; 3(1); 293–319; April 2016

[29] Stephanie M. Rizio; Ahmed Skali; "How often do dictators have positive economic effects? – Global evidence, 1858–2010"; The Leadership Quarterly; 31(3); 101302; June 2020

[30] Innovare: https://latin-dictionary.net/definition/23919/innovo-innovare-innovavi-innovatus; website visited on 18-4-24

[31] Ouroboros: https://nl.wikipedia.org/wiki/Ouroboros; website visited on 18-4-24

[32] Jan Verloop; Johan G. Wissema (Delft University NL); "Insight in Innovation – Managing Innovation by Understanding the Laws of Innovation"; ISBN 9780444516831; May 2004

[33] Jack Springman; "Drop innovation from your vocabulary"; Harvard Business Review; September 15, 2011: https://hbr.org/2011/09/drop-innovation-from-your-voca; website visited 18-4-24

[34] McKinsey Explainers "What is innovation?"; August 2022: chrome-extension:// efaidnbmnnnibpcajpcglclefindmkaj/ https://www.mckinsey.com/~/media/mckinsey/featured%20in sights/mckinsey%20explainers/what%20is%20innovation/what-isinnovation-final.pdf; website visited 18-4-24

[35] Jan Harmsen; Andre B. de Haan; Pieter L. J. Swinkels; "Product and Process Design"; ISBN 978-3-11-046772-7; e-ISBN (PDF) 978-3-11-046774-1; e-ISBN (EPUB) 978-3-11-046775-8; 2018

[36] Picture of First Oil: https://www.bp.com/en/global/corporate/who-we-are/our-history/first-oil.html; website visited on 18-4-24

[37] Picture of Olefin 3 naphtha cracker Sabic Geleen: https://iaprofessionals.nl/sabic-overweegt-sluiten-naftakraker/; website visited on 18-4-24

[38] David Bogle; Michael Fairweather; "Special issue – Computer Aided Process Engineering (CAPE) tools for a sustainable world"; Chemical Engineering Research and Design; 91(8); 1371–1372; August 2013; DOI:10.1016/j.cherd.2013.07.001

[39] https://apt.bci.tu-dortmund.de/professorship/team/prof-dr-ing-gerhard-schembecker/; website visited on 21-5-2024

[40] Prof.dr.ir. A. Kiss; Delft University, The Netherlands – personal communication

[41] Five Monkeys in a Cage: Intersol: https://intersol.ca/news/organizational-culture-and-the-5-monkeys-experiment/; visited on March 27th 2024

[42] Group Photo: a tool to create Dynamic interaction in a Team. First time application during a Self-Innovation Workshop, Apeldoorn, The Netherlands, February 2014

[43] Our Common Future; the World Commission on Environment and Development; October 1987; Oxford University Press; pages 16, 39, 41; ISBN 019282080X

[44] Kartikey Hari Gupta; "Sustainable Development Law – The Law for the Future"; 2016; Partridge Publishing India; ISBN 978-1-4828-7409-9

[45] "United Nations Sustainable Development Goals 2015": https://unglobalcompact.nl/sdgs/?utm_source=Adwords%2Fsdgs&utm_medium=Adwords%2Fsdgs&utm_campaign=Adwords%2Fsdgs&utm_id=cpc&utm_term=Adwords%2Fsdgs&utm_content=Adwords%2Fsdgs&gad_source=

1&gclid=CjwKCAjwh4-wBhB3EiwAeJsppE-jlvyVEbG9xFDiuUOF580BQLW4RsPLYdlnUA9h1YsRx_jXfaXz4
hoCf3MQAvD_BwE; visited on March 27, 2024

[46] SDG Action Awards: https://sdgactionawards.org/; visited on March 27, 2024

[47] https://sdgs.un.org/topics/industry; visited on March 27, 2024

[48] Jan Harmsen; Maarten Verkerk; "Process Intensification: Breakthrough in Design, Industrial Innovation Practices, and Education"; 2020; 237 pages; publisher de Gruyter; ISBN-10: 3110657341; ISBN-13: 978-3110657340; The Role of Sustainable Development Goals; 54–62.

[49] Andrzej Stankiewicz; T. Van Gerven, G. D. Stefanidis; "The Fundamentals of Process Intensification"; Intensified Reaction and Separation Systems, English, Weinheim, Germany, Wiley; 2019; 360 pages; ISBN (Electronic) 979-3-527-68015-3; ISBN (Print) 979-3-527-32783-6

[50] RAPID USA: https://www.energy.gov/eere/iedo/rapid-advancement-process-intensification-deployment-rapid-institute; website visited 12-4-24

[51] EFCE: https://efce.info/WP_PI.html; website visited 12-4-24

[52] Europic: https://europic-centre.eu/; website visited 12-4-24

[53] Frerich J.Keil; Process Intensification; Received September 8, 2017; accepted September 21, 2017; Rev. Chem. Eng. 2018; 34(2); 135–200; De Gruyter Publishers; https://doi.org/10.1515/revce-2017-0085

[54] European Roadmap for Process Intensification: chrome-extension://efaidnbmnnnibpcajpcglclefindmkaj/https://efce.info/efce_media/-p-531-EGOTEC-bc8c4b02f5191e7e8c5ff6095cffa126.pdf; page 22; website visited 12-4-24

[55] "Excel version of PI technologies, Process Functions, Driving forces and maturity level"; downloadable from the website of de Gruyter using the code that comes with a hard copy of the book

[56] Top 20 Engineering Companies to know: https://builtin.com/software-engineering-perspectives/engineering-companies; website visited on 25-4-24

[57] Best Companies for Chemical Engineers: https://www.zippia.com/chemical-engineer-jobs/best-companies-for-chemical-engineers/; website visited on 25-4-24

[58] Dog Material Balance: https://www.toothpastefordinner.com/index.php?date=080208; website visited on 25-4-24

[59] Achema 2024 website: https://www.achema.de/en/search; visited on 18-6-24

[60] Top Engineering Companies: https://builtin.com/software-engineering-perspectives/engineering-companies; website visited 15-5-24

[61] Largest Engineering Companies: https://companiesmarketcap.com/engineering/largest-companies-by-market-cap/; website visited 15-5-24

[62] Nadja Picard; Dr. Christoph Wallek; Daniela Wied; "Corporate Sustainability Reporting Directive (CSRD) – An analysis"; PriceWaterhouseCoopers GmbH, November 2023

[63] On FID: https://www.localcontent.com/what-is-a-final-investment-decision-fid/#:~:text=Final%20Investment%20Decision%20(FID)%20is,proceed%20with%20the%20project's%20execution; website visited on 27-6-24

[64] DACE; Price Booklet; Cost and Value; 36th edition, May 2023; publisher Bouwkosten.nl BV; ISBN 978-94-93312-21-0

Appendices

Appendix 1: client list

1	Afry	34	Lubrizol the Netherlands
2	Agro Knowledge Foundation	35	Lubrizol, Brecksville, OH, USA
3	(former) AkzoNobel industrial Chemicals B.V.	36	LyondellBasell Botlek
4	(former) AkzoNobel Surface Chemistry AB, Sweden	37	LyondellBasell Maasvlakte
5	(former) AkzoNobel Surface Chemistry, Herkenbosch	38	Meneba
6	Albemarle	39	Momentive
7	Allnex Resins B.V.	40	NedMag
8	Association for Farmers and Vegetable Growers	41	Netherlands Enterprise Agency
9	Berenschot	42	Neville Chemicals
10	Blue Terra	43	Norit
11	Carbogen Amcis Netherlands	44	Oostendorp Equipment Manufacturing
12	Chain Efficiency Foundation	45	Plasticon Europe B.V.
13	Chemtura Netherlands B.V.	46	PQ Europe
14	Cindu Chemicals B.V.	47	Private Equity Investors
15	Cosun	48	Rosier
16	CP Kelco	49	Royal Association of Dutch Chemical Industry
17	Delamine	50	Sabic Innovative Plastics
18	Dutch Institute for Fundamental Energy Research	51	Shell Chemicals
19	DMV International	52	Shin-Etsu PVC
20	DOW Chemical	53	Shin-Etsu VC B.V.
21	ECN BU Biomass, Coal and Environmental R&D	54	Smit & Zoon
22	ECN BU Energy Efficiency	55	Solar Netherlands
23	Eindhoven Institute for Renewable Energy Systems	56	TKI New Gas, NL
24	Emerald Performance Chemicals	57	TMC Chemical
25	Flowid	58	TNO Business Unit Oil & Gas
26	Fujifilm	59	University of Twente
27	Galvano	60	Undisclosed Oil Multinational, London
28	Honeywell UOP	61	Undisclosed Pharmaceutical Company, Belgium
29	Huntsman	62	Unimills
30	Impact	63	VitalFluid
31	Kemira Chemicals B.V.	64	Voyex
32	Kollo Silicon Carbide	65	VPR Refinery
33	Linde		

https://doi.org/10.1515/9783111383668-008

Appendix 2: intensified chemical technologies, including process functions, driving forces and maturity levels

Intensified technologies are listed in the columns on the left. Process functions, driving forces and maturity levels can be found in the top row from left to right. To make this readable, this Excel overview is converted into eight clusters of four pages. Each cluster categorizes about 11 technologies until all 81 technologies are categorized.

https://doi.org/10.1515/9783111383668-009

Technology	1 Absorbing	2 Adsorbing	3 Blowing	4 Breaking	5 Catalyzing	6 Centrifuging	7 Charging	8 Coagulating	9 Collapsing	10 Collecting	11 Combusting	12 Compressing	13 Condensing	14 Conducting	15 Convecting	16 Converting	17 Cooking	18 Cooling	19 Crushing	20 Crystallizing	21 Cutting
1 Adsorptive distillation		X															X				
2 Adv. HEX - plate														X				X			
3 Adv. HEX - spiral														X				X			
4 Centrifugal adsorption technology		X			X	X															
5 Centrifugal extractors						X															
6 Chemical looping					X			X						X	X						
7 Continuous oscillatory baffled reactors																					
8 Cryogenic separations														X	X	X	X	X			
9 Distillation - pervaporation systems																					
10 Dividing wall columns				X												X	X	X		X	
11 Ejector (Venturi) - based reactors				X				X				X					X	X			

2

Technology	Decharging (22)	Diffusing (23)	Disintegrating (24)	Disperging (25)	Dissolving (26)	Distilling (27)	Distributing (28)	Dividing (29)	Dripping (30)	Drying (31)	Emulsifying (32)	Evaporating (33)	Expanding (34)	Extracting (35)	Extruding (36)	Filtrating (37)	Flashing (38)	Foaming (39)	Freezing (40)	Gelling (41)	Heat transferring (42)	Heating (43)	Impulse transferring (44)	Isolating (45)
1 Adsorptive distillation						X						X					X							
2 Adv. HEX - plate																					X	X		
3 Adv. HEX - spiral																					X	X		
4 Centrifugal adsorption technology								X																
5 Centrifugal extractors		X		X							X			X									X	
6 Chemical looping																					X			
7 Continuous oscillatory baffled reactors		X												X							X			
8 Cryogenic separations						X		X				X					X				X			
9 Distillation - pervaporation systems						X		X				X					X				X			
10 Dividing wall columns						X		X				X					X				X			
11 Ejector (Venturi) - based reactors				X													X	X					X	

3

Technology	Process function																		Driving force - E			
	46 Liquefying	47 Mass transferring	48 Melting	49 Milling	50 Mixing	51 Permeating	52 Pouring	53 Pumping	54 Reacting	55 Separating	56 Shearing	57 Solidifying	58 Spinning	59 Spraying	60 Storing	61 Stripping	62 Sublimating	63 Wetting	1 Electric	2 Electromagnetic	3 Electrostatic	4 Free Gibbs
1 Adsorptive distillation										X												
2 Adv. HEX - plate																						
3 Adv. HEX - spiral																						
4 Centrifugal adsorption technology		X																				
5 Centrifugal extractors		X			X					X	X							X				X
6 Chemical looping		X							X	X	X		X									X
7 Continuous oscillatory baffled reactors		X			X				X	X	X											
8 Cryogenic separations		X								X												
9 Distillation - pervaporation systems		X				X				X												
10 Dividing wall columns	X									X												
11 Ejector (Venturi) - based reactors		X			X				X		X			X				X				X

4

Technology	Driving force - E field						Driving force - F field												Maturity / trl				
	5 Laser	6 Light visible	7 Microwave	8 Plasma	9 Temperature	10 Ultrasound	11 Affinity	12 Centrifugal	13 Concentration	14 Gravity	15 Impulse	16 Magnetic	17 Permeability	18 Pressure	19 Shear	20 Solubility	21 Velocity	22 Volatility	1	3	5	7	9
1 Adsorptive distillation					X		X											X			X		
2 Adv. HEX - plate					X																		X
3 Adv. HEX - spiral					X																		X
4 Centrifugal adsorption technology							X	X														X	
5 Centrifugal extractors								X			X				X	X						X	X
6 Chemical looping					X																X		X
7 Continuous oscillatory baffled reactors														X	X	X					X		X
8 Cryogenic separations					X		X		X		X					X		X					X
9 Distillation - pervaporation systems					X								X	X				X				X	
10 Dividing wall columns					X									X				X					X
11 Ejector (Venturi) - based reactors					X								X	X	X		X						X

Technology 5	1 Absorbing	2 Adsorbing	3 Blowing	4 Breaking	5 Catalyzing	6 Centrifuging	7 Charging	8 Coagulating	9 Collapsing	10 Collecting	11 Combusting	12 Compressing	13 Condensing	14 Conducting	15 Convecting	16 Converting	17 Cooking	18 Cooling	19 Crushing	20 Crystallizing	21 Cutting
12 Electric field-enhanced mixing							X	X		X				X							
13 Electric field-enhanced operations - other (e.g. fouling prevention)		X					X	X		X				X							
14 Electric field-enhanced extraction-dispersion		X					X	X		X											
15 Electric field-enhanced heat transfer		X					X							X	X			X			
16 Electrochemical reactors					X		X									X					
17 Extractive crystallization																				X	
18 Extractive distillation													X	X		X	X	X			
19 Foam reactors					X			X						X		X		X			
20 Gas-solid-solid trickle flow reactor	X	X			X											X					
21 Heat-integrated distillation columns	X												X				X	X			
22 Heat-integrated distillation	X												X				X	X			

6

Process function

Technology	22 Decharging	23 Diffusing	24 Disintegrating	25 Disperging	26 Dissolving	27 Distilling	28 Distributing	29 Dividing	30 Dripping	31 Drying	32 Emulsifying	33 Evaporating	34 Expanding	35 Extracting	36 Extruding	37 Filtrating	38 Flashing	39 Foaming	40 Freezing	41 Gelling	42 Heat transferring	43 Heating	44 Impulse transferring	45 Isolating
12 Electric field-enhanced mixing	X			X			X	X															X	
13 Electric field-enhanced operations - other (e.g. fouling prevention)	X		X	X			X																	
14 Electric field-enhanced extraction-dispersion	X		X	X			X				X			X										
15 Electric field-enhanced heat transfer																					X			
16 Electrochemical reactors	X																				X			
17 Extractive crystallization														X										
18 Extractive distillation						X		X				X		X			X				X			
19 Foam reactors		X		X				X								X		X			X			
20 Gas-solid-solid trickle flow reactor		X		X					X			X		X							X			
21 Heat-integrated distillation columns						X						X					X				X			
22 Heat-integrated distillation						X						X					X				X			

Technology	|	Process function																	|	Driving force - E			
		46 Liquefying	47 Mass transferring	48 Melting	49 Milling	50 Mixing	51 Permeating	52 Pouring	53 Pumping	54 Reacting	55 Separating	56 Shearing	57 Solidifying	58 Spinning	59 Spraying	60 Storing	61 Stripping	62 Sublimating	63 Wetting		1 Electric	2 Electromagnetic	3 Electrostatic	4 Free Gibbs
12 Electric field-enhanced mixing			X			X															X		X	
13 Electric field-enhanced operations - other (e.g. fouling prevention)			X								X										X		X	
14 Electric field-enhanced extraction-dispersion			X								X	X									X		X	
15 Electric field-enhanced heat transfer												X									X		X	
16 Electrochemical reactors			X				X			X	X										X		X	X
17 Extractive crystallization			X	X							X													
18 Extractive distillation		X	X								X													
19 Foam reactors			X			X				X	X													
20 Gas-solid-solid trickle flow reactor			X			X				X	X													
21 Heat-integrated distillation columns			X								X								X					X
22 Heat-integrated distillation			X								X								X					X

8 Technology	5 Laser	6 Light visible	7 Microwave	8 Plasma	9 Temperature	10 Ultrasound	11 Affinity	12 Centrifugal	13 Concentration	14 Gravity	15 Impulse	16 Magnetic	17 Permeability	18 Pressure	19 Shear	20 Solubility	21 Velocity	22 Volatility	1	3	5	7	9
Driving force - E							Driving force - force fields												Maturity / trl				
12 Electric field-enhanced mixing															X							X	
13 Electric field-enhanced operations - other (e.g. fouling prevention)																					X		
14 Electric field-enhanced extraction-dispersion																X					X	X	
15 Electric field-enhanced heat transfer																					X	X	
16 Electrochemical reactors									X							X							X
17 Extractive crystallization							X		X							X					X	X	
18 Extractive distillation					X		X		X							X		X					X
19 Foam reactors							X		X					X		X					X	X	
20 Gas-solid-solid trickle flow reactor					X		X		X							X							X
21 Heat-integrated distillation columns					X				X					X				X			X		
22 Heat-integrated distillation					X				X					X		X		X			X		

Technology 9	1 Absorbing	2 Adsorbing	3 Blowing	4 Breaking	5 Catalyzing	6 Centrifuging	7 Charging	8 Coagulating	9 Collapsing	10 Collecting	11 Combusting	12 Compressing	13 Condensing	14 Conducting	15 Convecting	16 Converting	17 Cooking	18 Cooling	19 Crushing	20 Crystallizing	21 Cutting
23 Hex reactor	X				X			X						X		X		X			
24 Hydrodynamic cavitation reactors					X			X				X				X	X	X			
25 Impinging streams reactor			X		X							X				X			X		
26 Induction – ohmic heating					X									X	X	X		X			
27 Ionic liquids	X				X																
28 Membrane absorption/stripping	X																				
29 Membrane adsorption		X																			
30 Membrane crystallization																					
31 Membrane distillation																				X	
32 Membrane extraction																X					
33 Membrane reactor (selective)					X																

10 Process function

	22 Decharging	23 Diffusing	24 Disintegrating	25 Disperging	26 Dissolving	27 Distilling	28 Distributing	29 Dividing	30 Dripping	31 Drying	32 Emulsifying	33 Evaporating	34 Expanding	35 Extracting	36 Extruding	37 Filtrating	38 Flashing	39 Foaming	40 Freezing	41 Gelling	42 Heat transferring	43 Heating	44 Impulse transferring	45 Isolating
23 Hex reactor																					X	X		
24 Hydrodynamic cavitation reactors				X																			X	
25 Impinging streams reactor		X		X																	X		X	
26 Induction - ohmic heating																					X	X		
27 Ionic liquids							X							X										
28 Membrane absorption/stripping		X						X				X				X								
29 Membrane adsorption		X						X								X								
30 Membrane crystallization		X						X								X								
31 Membrane distillation						X		X					X			X								
32 Membrane extraction								X						X		X								
33 Membrane reactor (selective)																X								

Technology 11	Process function																		Driving force - E			
	46 Liquefying	47 Mass transferring	48 Melting	49 Milling	50 Mixing	51 Permeating	52 Pouring	53 Pumping	54 Reacting	55 Separating	56 Shearing	57 Solidifying	58 Spinning	59 Spraying	60 Storing	61 Stripping	62 Sublimating	63 Wetting	1 Electric	2 Electromagnetic	3 Electrostatic	4 Free Gibbs
23 Hex reactor		X			X				X													X
24 Hydrodynamic cavitation reactors		X			X				X		X											X
25 Impinging streams reactor		X			X				X													X
26 Induction - ohmic heating		X							X											X		
27 Ionic liquids		X								X												
28 Membrane absorption/stripping		X				X				X						X						
29 Membrane adsorption		X				X				X												
30 Membrane crystallization												X										
31 Membrane distillation						X				X												
32 Membrane extraction		X				X				X												
33 Membrane reactor (selective)		X				X			X	X												X

Technology	Driving force - E						Driving force - force fields												Maturity / trl				
	5 Laser	6 Light visible	7 Microwave	8 Plasma	9 Temperature	10 Ultrasound	11 Affinity	12 Centrifugal	13 Concentration	14 Gravity	15 Impulse	16 Magnetic	17 Permeability	18 Pressure	19 Shear	20 Solubility	21 Velocity	22 Volatility	1	3	5	7	9
23 Hex reactor					X				X					X								X	X
24 Hydrodynamic cavitation reactors						X	X				X			X			X			X	X	X	
25 Impinging streams reactor					X				X		X				X					X	X		
26 Induction - ohmic heating					X																		X
27 Ionic liquids							X									X							X
28 Membrane absorption/stripping							X		X				X					X		X	X		
29 Membrane adsorption							X		X				X			X					X		
30 Membrane crystallization					X		X		X				X								X		
31 Membrane distillation					X		X		X				X	X				X					X
32 Membrane extraction							X		X							X						X	
33 Membrane reactor (selective)					X		X		X				X	X								X	

Technology 13

#	Technology	1 Absorbing	2 Adsorbing	3 Blowing	4 Breaking	5 Catalyzing	6 Centrifuging	7 Charging	8 Coagulating	9 Collapsing	10 Collecting	11 Combusting	12 Compressing	13 Condensing	14 Conducting	15 Convecting	16 Converting	17 Cooking	18 Cooling	19 Crushing	20 Crystallizing	21 Cutting
34	Membrane reactors (nonselective)					X											X					
35	Micro (channel) reactors					X											X		X			
36	Micro mixers																					
37	Microchannel heat exchangers																		X			
38	Microwave drying																					
39	Microwave heating																					
40	Microwave reactors - cat					X											X					
41	Microwave reactors - non cat																X					
42	Microwave reactors - polymer																X					
43	Microwave separation																					
44	Millisecond reactors					X											X					

14

Technology		Process function																								
		22	23	24	25	26	27	28	29	30	31	32	33	34	35	36	37	38	39	40	41	42	43	44	45
		Decharging	Diffusing	Disintegrating	Disperging	Dissolving	Distilling	Distributing	Dividing	Dripping	Drying	Emulsifying	Evaporating	Expanding	Extracting	Extruding	Filtrating	Flashing	Foaming	Freezing	Gelling	Heat transferring	Heating	Impulse transferring	Isolating
34	Membrane reactors (nonselective)																X								
35	Micro (channel) reactors																					X			
36	Micro mixers																								
37	Microchannel heat exchangers																					X	X		
38	Microwave drying										X											X			
39	Microwave heating																					X	X		
40	Microwave reactors - cat																					X	X		
41	Microwave reactors - non cat																					X	X		
42	Microwave reactors - polymer																					X	X		
43	Microwave separation								X						X							X			
44	Millisecond reactors																								

Technology 15	Free Gibbs (4)	Electrostatic (3)	Electromagnetic (2)	Electric (1)	Wetting (63)	Sublimating (62)	Stripping (61)	Storing (60)	Spraying (59)	Spinning (58)	Solidifying (57)	Shearing (56)	Separating (55)	Reacting (54)	Pumping (53)	Pouring (52)	Permeating (51)	Mixing (50)	Milling (49)	Melting (48)	Mass transferring (47)	Liquefying (46)
34 Membrane reactors (nonselective)	X												X	X			X				X	
35 Micro (channel) reactors	X													X				X			X	
36 Micro mixers																		X			X	
37 Microchannel heat exchangers																						
38 Microwave drying													X									
39 Microwave heating													X									
40 Microwave reactors - cat	X													X							X	
41 Microwave reactors - non cat	X													X							X	
42 Microwave reactors - polymer	X										X			X						X	X	
43 Microwave separation													X								X	
44 Millisecond reactors	X													X							X	

16

Technology	Driving force - E						Driving force - force fields												Maturity / trl				
	Laser (5)	Light visible (6)	Microwave (7)	Plasma (8)	Temperature (9)	Ultrasound (10)	Affinity (11)	Centrifugal (12)	Concentration (13)	Gravity (14)	Impulse (15)	Magnetic (16)	Permeability (17)	Pressure (18)	Shear (19)	Solubility (20)	Velocity (21)	Volatility (22)	1	3	5	7	9
34 Membrane reactors (nonselective)					X		X		X				X	X							X		
35 Micro (channel) reactors									X													X	X
36 Micro mixers														X									X
37 Microchannel heat exchangers					X																		X
38 Microwave drying			X																				X
39 Microwave heating			X																				X
40 Microwave reactors - cat			X						X												X		
41 Microwave reactors - non cat			X						X												X		
42 Microwave reactors - polymer			X				X		X						X					X			
43 Microwave separation			X																	X	X		
44 Millisecond reactors									X														X

Technology 17	1 Absorbing	2 Adsorbing	3 Blowing	4 Breaking	5 Catalyzing	6 Centrifuging	7 Charging	8 Coagulating	9 Collapsing	10 Collecting	11 Combusting	12 Compressing	13 Condensing	14 Conducting	15 Convecting	16 Converting	17 Cooking	18 Cooling	19 Crushing	20 Crystallizing	21 Cutting
45 Molecular-imprinted polymers		X								X											
46 Monolith reactors					X											X		X			
47 MPPE		X								X								X			
48 Multistream heat exchangers													X								
49 Other structured catalytic reactors (KATAPAK's, parallel pasage etc.)					X											X					
50 Pervaporation-assisted reactive distillation	X				X											X					
51 Photochemical reactors	X				X											X					
52 Plasma reactors							X									X					
53 Pulsed chromatografic reactors		X			X											X					
54 Pulsed combustion drying											X	X									
55 Pulsed compression reactor												X				X					

18

Technology	Decharging (22)	Diffusing (23)	Disintegrating (24)	Disperging (25)	Dissolving (26)	Distilling (27)	Distributing (28)	Dividing (29)	Dripping (30)	Drying (31)	Emulsifying (32)	Evaporating (33)	Expanding (34)	Extracting (35)	Extruding (36)	Filtrating (37)	Flashing (38)	Foaming (39)	Freezing (40)	Gelling (41)	Heat transferring (42)	Heating (43)	Impulse transferring (44)	Isolating (45)
45 Molecular-imprinted polymers																								
46 Monolith reactors																					X	X		
47 MPPE														X		X								X
48 Multistream heat exchangers																					X	X		
49 Other structured catalytic reactors (KATAPAK's, parallel pasage etc.)											X										X			
50 Pervaporation-assisted reactive distillation						X	X					X				X								
51 Photochemical reactors																								
52 Plasma reactors	X		X																		X	X		
53 Pulsed chromatografic reactors						X										X								
54 Pulsed combustion drying										X											X			
55 Pulsed compression reactor													X								X	X		

Technology 19	\[Process function\] 46 Liquefying	47 Mass transferring	48 Melting	49 Milling	50 Mixing	51 Permeating	52 Pouring	53 Pumping	54 Reacting	55 Separating	56 Shearing	57 Solidifying	58 Spinning	59 Spraying	60 Storing	61 Stripping	62 Sublimating	63 Wetting	\[Driving force - E\] 1 Electric	2 Electromagnetic	3 Electrostatic	4 Free Gibbs
45 Molecular-imprinted polymers		X																	X		X	
46 Monolith reactors		X							X	X												X
47 MPP		X							X	X									X			
48 Multistream heat exchangers																						
49 Other structured catalytic reactors (KATAPAK's, parallel pasage etc.)	X	X							X													X
50 Pervaporation-assisted reactive distillation		X				X			X	X									X		X	X
51 Photochemical reactors		X							X													X
52 Plasma reactors		X							X													X
53 Pulsed chromatografic reactors		X						X	X	X												X
54 Pulsed combustion drying		X			X			X	X													X
55 Pulsed compression reactor		X			X				X													X

Technology 20	Laser (5)	Light visible (6)	Microwave (7)	Plasma (8)	Temperature (9)	Ultrasound (10)	Affinity (11)	Centrifugal (12)	Concentration (13)	Gravity (14)	Impulse (15)	Magnetic (16)	Permeability (17)	Pressure (18)	Shear (19)	Solubility (20)	Velocity (21)	Volatility (22)	trl 1	trl 3	trl 5	trl 7	trl 9
45 Molecular-imprinted polymers					X		X																X
46 Monolith reactors					X																X	X	X
47 MPP					X		X		X							X							X
48 Multistream heat exchangers					X																		X
49 Other structured catalytic reactors (KATAPAK's, parallel pasage etc.)									X														X
50 Pervaporation-assisted reactive distillation							X		X									X				X	X
51 Photochemical reactors		X					X													X	X		
52 Plasma reactors				X			X														X	X	X
53 Pulsed chromatografic reactors							X		X							X				X	X		
54 Pulsed combustion drying					X									X							X	X	X
55 Pulsed compression reactor					X									X						X	X		X

Technology 21	1 Absorbing	2 Adsorbing	3 Blowing	4 Breaking	5 Catalyzing	6 Centrifuging	7 Charging	8 Coagulating	9 Collapsing	10 Collecting	11 Combusting	12 Compressing	13 Condensing	14 Conducting	15 Convecting	16 Converting	17 Cooking	18 Cooling	19 Crushing	20 Crystallizing	21 Cutting
56 Pulsing operation of multiphase reactors	X	X			X							X				X					
57 Reactive absorption	X				X											X					
58 Reactive comminution		X		X												X			X		
59 Reactive condensation								X					X			X					
60 Reactive crystallization / precipitation					X											X				X	
61 Reactive distillation					X								X			X					
62 Reactive extraction																X					
63 Reactive extrusion					X											X		X			
64 Reverse flow reactors					X							X				X					
65 Rotating annular chromatographic reactor		X			X											X					
66 Rotating packed beds	X	X			X	X									X	X					

22

Technology	Process function																							
	22 Decharging	23 Diffusing	24 Disintegrating	25 Disperging	26 Dissolving	27 Distilling	28 Distributing	29 Dividing	30 Dripping	31 Drying	32 Emulsifying	33 Evaporating	34 Expanding	35 Extracting	36 Extruding	37 Filtrating	38 Flashing	39 Foaming	40 Freezing	41 Gelling	42 Heat transferring	43 Heating	44 Impulse transferring	45 Isolating
56 Pulsing operation of multiphase reactors				X					X			X	X					X			X			
57 Reactive absorption					X								X											
58 Reactive comminution																								
59 Reactive condensation																					X			
60 Reactive crystallization / precipitation																								
61 Reactive distillation						X						X									X			
62 Reactive extraction					X									X										
63 Reactive extrusion															X						X			
64 Reverse flow reactors																					X	X		
65 Rotating annular chromatographic reactor								X																
66 Rotating packed beds																					X		X	

Technology 23

#	Technology	46 Liquefying	47 Mass transferring	48 Melting	49 Milling	50 Mixing	51 Permeating	52 Pouring	53 Pumping	54 Reacting	55 Separating	56 Shearing	57 Solidifying	58 Spinning	59 Spraying	60 Storing	61 Stripping	62 Sublimating	63 Wetting	1 Electric	2 Electromagnetic	3 Electrostatic	4 Free Gibbs
56	Pulsing operation of multiphase reactors		X			X				X	X								X				X
57	Reactive absorption		X							X	X												X
58	Reactive comminution				X					X		X											X
59	Reactive condensation	X	X							X	X												X
60	Reactive crystallization / precipitation		X							X	X		X										X
61	Reactive distillation		X							X	X												X
62	Reactive extraction		X							X													X
63	Reactive extrusion		X	X						X													X
64	Reverse flow reactors		X						X	X	X												X
65	Rotating annular chromatographic reactor		X							X	X												X
66	Rotating packed beds		X			X				X	X												X

24

Technology	Laser 5	Light visible 6	Microwave 7	Plasma 8	Temperature 9	Ultrasound 10	Affinity 11	Centrifugal 12	Concentration 13	Gravity 14	Impulse 15	Magnetic 16	Permeability 17	Pressure 18	Shear 19	Solubility 20	Velocity 21	Volatility 22	1	3	5	7	9
	Driving force - E						Driving force - force fields												Maturity / trl				
56 Pulsing operation of multiphase reactors					X				X					X									X
57 Reactive absorption									X							X						X	
58 Reactive comminution											X				X					X	X		
59 Reactive condensation					X				X									X			X	X	X
60 Reactive crystallization / precipitation									X						X	X					X	X	X
61 Reactive distillation					X				X							X		X					X
62 Reactive extraction					X		X									X							X
63 Reactive extrusion											X												X
64 Reverse flow reactors					X				X														X
65 Rotating annular chromatographic reactor							X									X				X	X		
66 Rotating packed beds							X	X	X		X					X		X			X	X	X

Technology 25	1 Absorbing	2 Adsorbing	3 Blowing	4 Breaking	5 Catalyzing	6 Centrifuging	7 Charging	8 Coagulating	9 Collapsing	10 Collecting	11 Combusting	12 Compressing	13 Condensing	14 Conducting	15 Convecting	16 Converting	17 Cooking	18 Cooling	19 Crushing	20 Crystallizing	21 Cutting
67 Rotor-stator mixers						X															
68 Simulated moving bed reactor		X			X											X					
69 Sonochemical reactors									X			X				X					
70 Spinning disk reactors (SDRs)	X	X			X	X		X						X		X		X			
71 Static mixers																					
72 Static mixers-heat exchangers																					
73 Static mixers-reactors															X	X					
74 Structured internals for mass transfer operations		X																			
75 Supercritical reactions	X														X	X					
76 Supercritical separation	X														X						
77 Supersonic gas-liquid reactors			X												X	X					

26

Technology		Process function																								
		22 Decharging	23 Diffusing	24 Disintegrating	25 Disperging	26 Dissolving	27 Distilling	28 Distributing	29 Dividing	30 Dripping	31 Drying	32 Emulsifying	33 Evaporating	34 Expanding	35 Extracting	36 Extruding	37 Filtrating	38 Flashing	39 Foaming	40 Freezing	41 Gelling	42 Heat transferring	43 Heating	44 Impulse transferring	45 Isolating
67	Rotor-stator mixers				X																			X	
68	Simulated moving bed reactor														X										
69	Sonochemical reactors			X		X																	X		
70	Spinning disk reactors (SDRs)				X	X																X	X	X	
71	Static mixers																							X	
72	Static mixers-heat exchangers																					X	X	X	
73	Static mixers-reactors																					X	X		
74	Structured internals for mass transfer operations							X																	
75	Supercritical reactions				X	X																			
76	Supercritical separation														X										
77	Supersonic gas-liquid reactors				X	X																		X	

Technology 27	Driving force - E				Process function																		
	1 Electric	2 Electromagnetic	3 Electrostatic	4 Free Gibbs	46 Liquefying	47 Mass transferring	48 Melting	49 Milling	50 Mixing	51 Permeating	52 Pouring	53 Pumping	54 Reacting	55 Separating	56 Shearing	57 Solidifying	58 Spinning	59 Spraying	60 Storing	61 Stripping	62 Sublimating	63 Wetting
67 Rotor-stator mixers						X			X													
68 Simulated moving bed reactor			X	X									X									
69 Sonochemical reactors						X			X				X									
70 Spinning disk reactors (SDRs)				X		X			X				X	X	X							
71 Static mixers									X													
72 Static mixers-heat exchangers									X													
73 Static mixers-reactors				X		X			X				X									
74 Structured internals for mass transfer operations						X																
75 Supercritical reactions				X		X			X				X									
76 Supercritical separation			X			X			X					X								X
77 Supersonic gas-liquid reactors				X		X			X				X									

28

#	Technology	5 Laser	6 Light visible	7 Microwave	8 Plasma	9 Temperature	10 Ultrasound	11 Affinity	12 Centrifugal	13 Concentration	14 Gravity	15 Impulse	16 Magnetic	17 Permeability	18 Pressure	19 Shear	20 Solubility	21 Velocity	22 Volatility	trl 1	trl 3	trl 5	trl 7	trl 9
				Driving force - E							Driving force - force fields											Maturity / trl		
67	Rotor-stator mixers								X			X				X								X
68	Simulated moving bed reactor					X		X													X	X	X	X
69	Sonochemical reactors						X														X	X	X	X
70	Spinning disk reactors (SDRs)					X			X	X					X	X							X	
71	Static mixers														X	X								X
72	Static mixers-heat exchangers					X										X								X
73	Static mixers-reactors					X									X									X
74	Structured internals for mass transfer operations							X							X	X								X
75	Supercritical reactions					X		X		X							X		X			X	X	X
76	Supercritical separation					X		X		X							X		X			X	X	X
77	Supersonic gas-liquid reactors														X	X								X

Technology **29**	1 Absorbing	2 Adsorbing	3 Blowing	4 Breaking	5 Catalyzing	6 Centrifuging	7 Charging	8 Coagulating	9 Collapsing	10 Collecting	11 Combusting	12 Compressing	13 Condensing	14 Conducting	15 Convecting	16 Converting	17 Cooking	18 Cooling	19 Crushing	20 Crystallizing	21 Cutting
78 Supersonic gas-solid reactors			X									X									
79 Ultrasound-enhanced crystallization															X	X					
80 Ultrasound-enhanced phase dispersion / mass transfer														X							
81 Viscous heating																		X		X	

Technology **30**	Process function																							
	22 Decharging	23 Diffusing	24 Disintegrating	25 Disperging	26 Dissolving	27 Distilling	28 Distributing	29 Dividing	30 Dripping	31 Drying	32 Emulsifying	33 Evaporating	34 Expanding	35 Extracting	36 Extruding	37 Filtrating	38 Flashing	39 Foaming	40 Freezing	41 Gelling	42 Heat transferring	43 Heating	44 Impulse transferring	45 Isolating
78 Supersonic gas-solid reactors				X	X		X																X	
79 Ultrasound-enhanced crystallization					X																			
80 Ultrasound-enhanced phase dispersion / mass transfer				X																	X			
81 Viscous heating																					X	X		

Technology 31	Process function																		Driving force - E			
	46 Liquefying	47 Mass transferring	48 Melting	49 Milling	50 Mixing	51 Permeating	52 Pouring	53 Pumping	54 Reacting	55 Separating	56 Shearing	57 Solidifying	58 Spinning	59 Spraying	60 Storing	61 Stripping	62 Sublimating	63 Wetting	1 Electric	2 Electromagnetic	3 Electrostatic	4 Free Gibbs
78 Supersonic gas-solid reactors		X			X				X													X
79 Ultrasound-enhanced crystallization		X								X		X										
80 Ultrasound-enhanced phase dispersion / mass transfer		X																				
81 Viscous heating													X									

Technology **32**	Driving force - E						Driving force - force fields												Maturity / trl				
	5 Laser	6 Light visible	7 Microwave	8 Plasma	9 Temperature	10 Ultrasound	11 Affinity	12 Centrifugal	13 Concentration	14 Gravity	15 Impulse	16 Magnetic	17 Permeability	18 Pressure	19 Shear	20 Solubility	21 Velocity	22 Volatility	1	3	5	7	9
78 Supersonic gas-solid reactors															X								X
79 Ultrasound-enhanced crystallization						X			X							X	X						X
80 Ultrasound-enhanced phase dispersion / mass transfer						X														X	X	X	
81 Viscous heating															X								X

Index

https://doi.org/10.1515/9783111383668-010

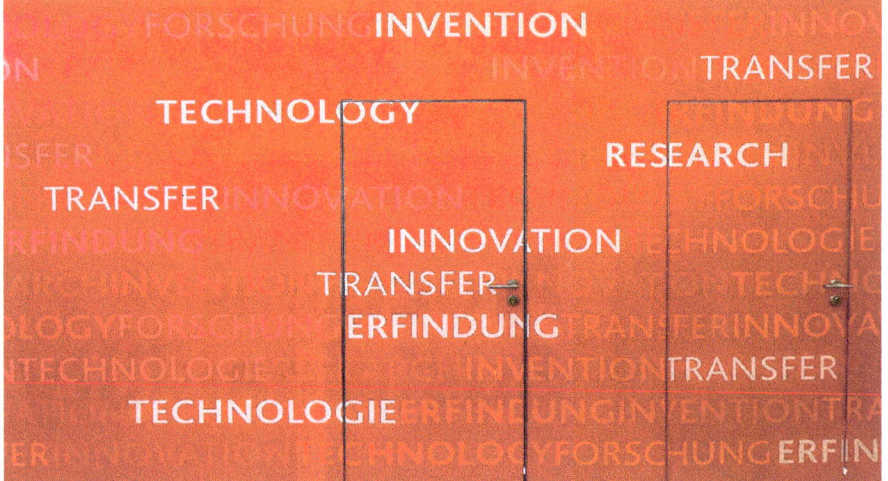

www.ingramcontent.com/pod-product-compliance
Lightning Source LLC
Chambersburg PA
CBHW081538220326
41598CB00036B/6480